The Time-Space Traveler's Handbook

By

Robert B Cronkhite

Table of Contents

Initiating

Many have wondered how far our experimentations have gone in pursuance of the explorations based on Hypergeometry and its interpretations of time and space. The 'what ifs' and their implications are profound. Practicing theory as mechanics is when and where the difficult, but provocative, surfaces. Some will consider this science fiction if it comforts them; but you my dear reader, dare to take the ride and test all of this.

It is imperative to realize that any actions ... within time or within space ... affect a reaction within the other for any object. In the present or 'now' of an object (as repeatedly related in my other works) it is kinetically and inertially above the background noise signal ratio. What this means is that it is inertially, geometrically related to other objects at that present now time so as to mechanically inter-react with them. For example one cannot pass through a wall, for a door is needed to go from one room to another. If an object is either past or future in relation to a frame of the inertial reference of now, then the wall and the person are kinetically irrelevant to each other.

If however, an object or a person is in a far enough different time or space so as not to be kinetically or inertially relevant to the other, then it has no relevant kinetic interference to the other. There is always some 'below the noise shadow' though, as we understand Maxwell's equations that relate time for a moment to space or space to a time of enough distance. So then a wall in London has no bounds upon a person in New York at the same time, because of spatial distance. Thus, temporally, a wall present in 1895, if torn down afterward will have no bounds upon a person in that exact same space in 2010.

In the treatise that follows are the implications from the results of further experiments of mine since the publishing of HYPERSPHERE … A JOURNEY AT THE SPEED OF GEOMETRY (original and revised versions) and ODYSSEY OF THE AGES. It expands upon the further use of the Temporal Diffraction Grating and the Hyperplane, along with the Hypersphere Discs. I define these as being complements to cameras and sound recording devices, considered 'Quantum Machines'. Others involved agree that as thus, they have a far different perspective upon us as humans and the universe around us.

If some consider me and others of the same mind, ahead of our time, too counterintuitive or just plain fanciful, it is because we have not felt restrained to paradigms of many physicists. We have based everything on established science and math of the eighteenth through the twenty-first centuries. Any one is welcome to attempt these things as able. Some have hoped to prove me wrong, but the years are going by and I await their scientific disapproval. Meanwhile, "Welcome to the future, and the past and the present!"

Robert B Cronkhite, 2014 AD

Chapter One

Photographs
Not Typically Taken

Audio-visual recording of something in one place and then in a different place (space) is usually an easy process, for all things at the same present (now) moment are kinetically and inertially able to interfere. For instance, you can touch a flower or go through a doorway and you can take a photograph and/or make a sound recording of such.

To record photons is easier than making a copy of any acoustic-mechanical emanations, better known as sound. Certain conditions need to be attained to record sound. The photons of light have far more momentum and effect than the sound pressure propagating for the particular event at hand. The light is faster and has much more inertial/kinetic interference upon the detectors, as film in a camera contrasted with sound reaching a microphone. Also recording an image in color is harder than black and white, and capturing motion may be more difficult. In these situations, there is very little interference from the present time of the experiment upon the time of the past interest. Future is more difficult than past also; the future has a far weaker signal. A lot of what James Clerk Maxwell wrote about is very fundamental to all this. Photons from the past 'leak' the best, is one way to state this for better understanding.

One bit of digression, if you would allow me: we are all already traveling through time and space. It is at present our environment of experience, but so few tend to think this way. I will further explain this at a later time. It will be important to consider along with any interest in past or future time, the space involved as well. Time and space are linked, and one needs the other. For example to consider a traverse from 2010 to 1895 in time, one has to also consider a traverse of 115 light years in space. The speed of light is the fastest in the inertial frame of reference that any information may flow. To be able to kinetically and inertially change one's kinetic-inertial frame of reference, one needs to acquire a Non-inertial state (tachyon) at 370,000 times the speed of light (c). This is what we understand as the tachyons 'rest' speed in relation to us.

Included in this guide will be photographs and explanations of things more as we go along. This is not meant to be a large cumbersome book, but rather a convenient manual for those 'Time-Space Travelers' who would seriously attempt these things. The fascination is strong enough to encourage and challenge those who would take the time to contemplate and build such experiments, to see just how far one might go.

My interest in 1895 is because of H G Wells' timeless work, *THE TIME MACHINE* and the existence of my great-great-great grandfather and his bakery-confectionary at that particular time. My research and endeavor to connect events and times between 2010 and 1895 was very personal to me. I knew when, as well as where, his store was located. I had to consider outsiders trying to dissuade my experiments, without allowing myself to be hindered by opinions and assumptions, but rather only mathematics and science. This meant that I was not limited by any corporate or politically biased educational institution worried about funding or profit. I had the fire in me to keep things to myself and indulge as a hobby such exploits. If things did not work out, then no one would know the better. If things did become interesting, then I would take notes, share such carefully and in time maybe publicize some of what I felt would be important.

Unfortunately;, so much herein leads to the possibilities of misuse that I have had to be very careful of what I would share. Our civilization is not always very civilized. I am of the opinion that we really cannot govern ourselves and actually do not have the wisdom to handle the increasing volume of knowledge, nor its speed.

Those living in the nineteenth century so looked forward with great interest and expectations to the future, and now presently many of us in the twenty-first century look longingly back to the past; each people dreamed of a better world to come, or of one nostalgically missed.

I shall purposely include repeatedly herein, concepts and definitions from my aforementioned writings. The reason is so much like quantum mechanics and relativity … so counterintuitive. Our normal and understood experiences are classical and make sense for most of humanity's existence. But with the atomic-age and the space-age, many ideas are not easily understood or accepted. Neither do I have the final stage of understanding. It is all to me a wonder that even glimpses are possible. To be able with much effort to go even farther is more than at first was considered attainable. Here established science is fine. But with unlimited personal time and initiative, one can, without the previously mentioned constraints, not worry if tenure will be affected or career side-lined by such an endeavor.

As with the early stages of astronomy, the automobile and airplane, atomic theory, radio and the personal computer, this is a field of much promise for the amateur scientist who has the greatest advantage to pursue such research. Here one has to tediously attempt for extremely high resolution and seek 'shadow between the shadows, and silence between the silences'. As you may also have wondered about the possibilities of traveling in not only space, but also in time, you can almost feel that theory has been leading into something more substantial. You find bits and pieces, and must jump to conclusions perhaps not yet or ever, fully provable. It is the same for astrobiology and evolution. Such precarious endeavors of study, much like detective work, have implications; we're tempted to try to run with it and see where it goes. In some, possibly many cases, such are still not provable, but we nevertheless have the fire to try to find out after all what is going on.

In the end, theory becomes fact solely with repeatable and verifiable evidence. In the interim, even though we may not understand, it is quite a thrilling ride. Many outside of this study are still trying to prove by implication what they subjectively feel to be some glimpse of a greater reality. In some things we may never really understand for a long time, but that does not hinder each of us to promote upon others what we want to believe. Here for the first time, is the most quantifiable of research into the possibilities of not only space-traveling, but also of time-travel.

The Temporal Diffraction Grating was the first Hypergeometric machine to be utilized. With the use of a laser, it would unravel the flow of time forward, even when spatially the lasers were going in opposite directions. Sounds confusing? Let me illustrate more historically.

Upon opposite sides of the Earth along the terminator, it is dawn on one side and dusk on the other. The stars are shining their light from the past. Yet to each observer the rays of light from the past are traveling in opposite spatial directions. They both, from either side, see into the past and the only difference is that the light is spatially opposite, but the same direction temporally. Such simplicity once one acquires a different, more conceptually lucid view. Therefore, temporally they are both looking into the past, and yet in the same direction.

My first illustrations are two photographs, both taken in 1895, but from opposite directions through the same axis. I dare anyone to try to take these same photographs now. It was late spring, early summer. One view was taken from the left, second floor window of the Simpson Building on 27

Market Street in Amsterdam, NY. The other, looking back, was from the opposite side of the street with its perspective of view in almost 180 degree opposition from the first, but some hours later, as the Sun had coursed farther to the west in the sky of that time.

The camera was already considered a quantum machine, but enabled by the discs, inherent with such photon detection. What information that was considered just fable and shadow, is now recorded upon the CD within and able to be re-presented. The opposition ability of a camera is because of geometry during entanglement of the oscillations of the parasitic of the Hypersphere driving the discs, and their harmonic reflection of such entanglement. The disc has to be in close proximity of the camera, or any recording device, as well as the one recording, if they want to experience such. The 'travel' time to 1895 from 2010 is almost three hours, and the velocity to get there is 370,000 times the speed of light in reverse time.

In the few hours to translate (2X) the geometry of 180°, as each geometric translation is equal to 90°, one sees that the Sun has moved in the sky and that the tarp upon the pipe structure is now removed.

For these images, no inertial or kinetic interference has been known to be initiated. All we have done was to amplify what was already such an extremely week signal. Thus the universe as we know it is in some ways recorded, or better leaving trails that can be amplified. It is written that we have a great cloud of witnesses, and so it is here we have something to ponder. Scientifically there is no privacy, so all would then be recorded or detectable, given enough resolution of sensitivity. When one oscillates the Hypersphere and has proper articulation of a disc, then any camera is a grand window of not only space, but of time. Fascinating to be able to observe and measure the reality of human governments' actual activities, as opposed to the propaganda of such empire building machines.

Later we shall focus upon the discs that accompany a Hypersphere. Time and space will be better understood to be very inter-related as we progress. Though bear with me as I, in addendum, bring in scientific principles while in this writing try to keep all practical mechanics for the amateur hobbyist, until those in power disdain such flow of information upon the common. One needs to remember, with such approach to history, one is not limited by an elite's self-promoting dithers. No, the People can actually perceive more of what has happened, so to better govern themselves, if they have the maturity to do so.

The simple production of the discs easily allows one to parasitically 'ride' upon the local activity of a larger and far more complex Hypersphere. It is the 'poor man's time-space machine'. And because of Non-inertial and kinetic interference, it is actually a reference device without the complications of causality paradoxes. The world line, just more detectable, that is always there but so weak to measure and observe. In those one cannot change the past.

The discs are plainly parasitic oscillators, but their local effect upon recording devices is where they are so influential, as well as so inexpensive. Each disc is about 10 to 13 cm in diameter. Within are two mass spheres that oscillate in harmony or resonance to the live Hypersphere. The masses are connected by a spring, and all they do is vibrate in response to the aforementioned driving of the local Hypersphere. Yes, one will need to build the prime oscillator, the Hypersphere as well. Or if such is readily available, perhaps used by the government or an institution, then you may get the 'free ride' I had mentioned before.

Let us try to concentrate upon the discs required and thus have the most economical and simplest start. One disc is all that is required, though two or more can give interference pattern applications. But with one disc, you have a 'window' of temporal access that can now be amplified, that has always been there.

Reviewing the Temporal Diffraction Grating, one can see that the discs are really elaborations of the detectors used originally. These detectors were merely photo diodes that were cooled to liquid nitrogen temperatures, nothing more. With the inclusion of mechanical detection of the fluctuations in the local Inertial Geometric and the variations in the same local space-time curvature of any moving mass, one can also detect these fluctuations because of the parasitic effect upon other masses around. These are very resonant when the masses sensing these waves of fluctuation are also of a harmonic wavelength of diameter, especially as spheres. It seems when the discs are in resonance, they too generate a second fluctuation at the incident frequency and in harmonics as well. Now theory begs more, but with still continuing attempts in trying to understand, the best is just repeated experiments as with a camera. Over the time of these different explorations, we were able to collect interesting artifacts to further put our hypothetical thoughts around. We admit not all is clear to our understanding. Perhaps you will further such research. We at this writing have gone quite far and still are continuing experiments, while yet only glimpsing so little of such greater things.

You can see from the photograph of a disc in the palm of my hand, that it is not very large or encumbering at all. Herein is a subset of the Hypersphere. Again all it does is resonate with the Hypersphere's fluctuations of the local space-time curvature, or better the local Inertial Geometric. When a camera or any recording device is placed near enough to it, then some very interesting footages are possible.

I will use an illustrations of the Hyperplane and the Temporal Diffraction Grating. Both of these are 'Third Level Machines' which correspond to the coefficient of the World Line Formula: $ds^2 = dx^2 + dy^2 + dz^2 - dt^2$. When the coefficient is positive, it is space-like; when it is zero, it is light-like; and when the coefficient is negative, it is time-like. Thus the third coefficient is the Third Level Machines. Why? Because the space-like machines are what we mostly have experienced. Likewise with the invention of radio and light emitting devices, we have the light-like to Second Level Machines. Now with the realm of the most difficult, the negative coefficient and time-like are thus Third Level Machines.

All extreme relativistic and quantum devices are Third Level Machines for they are predominately time-like. When something is predominately time-like, then all other conditions such as space-like and light-like are the minors. For most of human history, space-like have been predominate, whilst the other two factors were very minor. They are always there, but most of the time so very slight to be almost irrelevant. All machines minor the other two factors, so light-like has space-like and time-like as minors in function. They are there and affected, but slight enough to not be of any obvious consequence.

Now I would like to explain the second machine to be developed. First was the Temporal Diffraction Grating, then the Hyperplane, consequently the Hypersphere with its accompanying discs. But let us save the first one for last. Next I will illustrate the Hyperplane.

This device has a rotating laser beam and within the inner radius less than light, all functions were in forward duration detected. We ran a series of digital or binary numbers that would track correctly the forward sequence. According to the following illustration, this would be the inner detector.

At the white rim was another detector where the sequence was frozen in time, and outside this ring, the outer detector recorded the sequence in reverse. Thus a photonics Event Horizon occurred at the white rim which equaled the speed of light for the rotation of the beam's radial velocity. Inside the white rim, the beam was below the speed of light, and outside the white rim the rotational velocity was faster than light … so the sequence was reversed.

The Hyperplane was the second of these Third Level or Hypergeometric Machines, which is very useful to observe and measure, and so quantify time. Quantification began with the Temporal Diffraction Grating. Previously, time was considered esoteric and unquantified as a function. All it was given was an abstract consideration, while other phenomena were looked at as concrete events. Now with actual situations with which to test the flow of time and some sequence of events, it was possible to better understand that part of our universe (or superuniverse of universes and some even say we should call it our 'Omniverse') is far more than just a spatial entity.

This is confounding for many, for just like gravity, time is also miss-defined. Once more, as stated in my past books, time is treated as a very substantial part of the world around us.

Considering as Inertial Geometry, gravitation and time are more influential than just abstractions, we suddenly have more of the reality around us than before. Take into consideration the November, 2014 British Interplanetary Society's magazine *Spaceflight's* article, 'The Tachyon World', and we have a substrate to our superuniverse where our local universe floats upon a surface equal to the speed of light. Fascinating are the implications of all this, that now a new perspective of time and the sequences it entails, so much seem to be less counterintuitive and sensible.

The Hyperplane was a meter in diameter when operating very fast by use of mirrors and few mechanical parts for such a radius. The longer was better with a diameter of a kilometer and far more resolute. But even the smaller version sufficed well enough. I spun a laser beam in a radial pattern at such a speed that at some point, the radial velocity equals the speed of light. The beam of the Hyperplane as it rotated faster than light, and sequences in reverse.

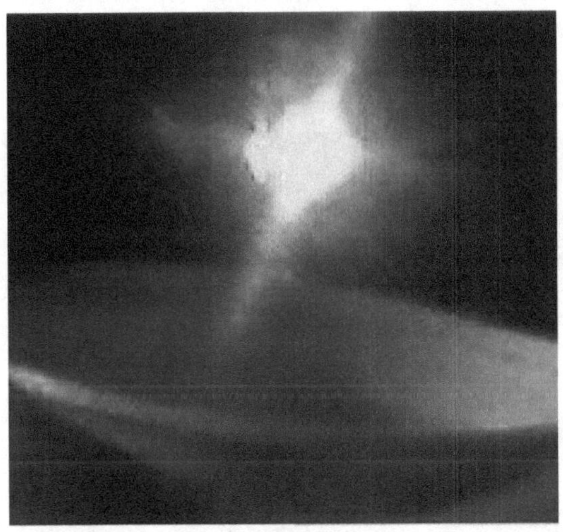

Now we come to the Temporal Diffraction Grating, which was the first Third Level Machine developed. It was the first successful attempt to quantify time in a holistic way. In other words, one could observe the measuring of a laser beam in past, present and future ... all at the same time. It spatialized that which was temporal. It was an equalizer of sorts, as this illustration of Temporal Diffraction Grating with past, present and future spatially indicated

In the next illustration, the red laser beam flows from left to right. The second vertical and longest white line is the present and at the diffraction grating, where the measurement of 'now' occurs. This is discerned in that it is the only place recognizable where the indication of an event has occurred and such is the diffraction itself by photodetectors. It is the longest of the vertical white lines for 'now', the strongest kinetic/inertial signal we can record, and also the easiest. The white lines represent the relative amounts of energy of past, present and future. The left vertical line represents past, and the photodetector records

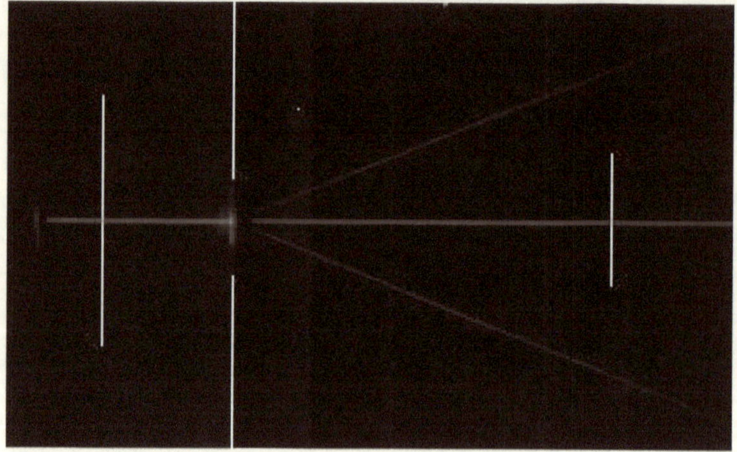

this first, for it occurs before our designated 'now' and is less in measured energy as it dissipates in relation to the 'now' signal. The future signal is the least strong and requires Cryogenics to enable the photodetector to sense the future signal at the same time the other signals are sensed. All these signals occur in temporal forward sequence obviously. In reality, it is a 'photon clock'.

When each photodetector is measuring, isolated from the others, no cryogenics is needed. But to measure this flow of time at the same time, implies a second dimension of time: therefore the greatest in strength is 'now', then the 'past' and least in strength the 'future'. It is a counterintuitive experiment we called the 'Where is 'now' experiment', which had grand implications for time. It is thought there is an 'Absolute Time' to reference what we experience as the flow of normal time. And the question is begged, if we are glimpsing this Absolute Time (or to honor Sir Isaac Newton, this 'Newtonian Time') then to what are we referencing these two time dimensions? Well, it seems to be at first a hall of mirrors, but when thought through we had perhaps a glimpse of Newtonian Time shadowing whilst in measuring our normal flow of time.

All this was required first to be able to progress on to the Hyperplane, then the Hypersphere and the accompanying discs. Each step helped us to better acquire a sense of time away from the dogmatic scrutiny of the limited investigation of established science before all this. Up to this time, very little from serious university or government studies existed. In most cases, theoretical work was all that was done and not much really serious undertakings in experimentation. The reason was that practicality was not considered, for time was so thought to be non-substance and not for quick profit of any industry, especially defense of mass production for practical consumerism.

This is where we had the advantage. Without restrictions from any aforementioned bureaucracy, we had a wide field to play in. You, as an amateur, do as well.

The narrow-mindedness of the present state of science, which is difficult to accept since we have broad-mindedness in so many other areas, has hidden by paradigm such glimpses from the established academia. Granted, one has to go out on a limb to discover new things, and even in today's scientific environment, we find it hard to believe we are limited in some areas where we are required to follow dogma, rather than repeatable verifiable evidence. Real science is as unpopular among 'scientific' institutions in many cases as it was in the dark ages of past human history. The modern Giordano Bruno, Galileo Galilei and Nicolaus Copernicus had to suffer the same. Try to promote one's repeatable and verifiable scientific evidence against the neo-modern dark ages, and you too will be 'burned at the stake' or lose your tenure.

Instead of venting frustrations, just give me fair credit and build these things. Try for yourself to see where your version of these experiments lead. Again, test these things as to whether they are science fiction or far more. Don't lean upon the opinions of some academics who pass judgement before objective research is done. You who are reading this already have realized since your childhood that you were different and are used to it, though it sometimes breaks your heart. The church was so often blamed for such, whilst the religion of dogmatic academics also throws the same stones.

The Temporal Diffraction Grating was about two meters in length and was the simplest of the three machines. The difference in these machines was that for the first time geometry, especially Hypergeometry, was finally mechanized. I am including the two discs as part of the Hypersphere. For without the Hypersphere, they are not complete in themselves.

With this one can take a transparent photograph and scan the laser through it. Remember, that for every nanosecond of time from the laser aperture to the future equals about 33 cm. So looking down over the apparatus or from any angle from the axial line of the linear laser beam, one can take in by perspective, past, present 'now' and future. Here we define now with our point of measurement and observation of diffraction. It is the event we perceive from the normal course of the laser beam. It is, as they ask, 'what's happening'.

With the other detectors we can thus choose our 'now' from anytime in the availed past, present or future, or even have multiple 'nows' to observe from any angle away from the beam's linear flow. It is as though all that is happening along the laser beam is its world line, and we can almost Non-inertially or Non-kinetically experience the series of events along this world line of the Temporal Diffraction Grating.

This simple machine is nonetheless profound. It is said that 'oft times the most profound begins with the most subtle' and here it is dramatically so. On top of your kitchen table or where ever your homemade laboratory is located with such a pittance of economics, you are observing what very expensive institutions of governments and universities are missing. This is where I can appreciate the expression I have so often stated, 'measuring the shadow between the shadows, or the silence between the silences'. Not only measuring, but respectfully seeing and hearing, if done properly.

I attempt to say this with the limited tongue of human language, trying to poetically express what is actually difficult to describe. It is similar to having exact numbers that make it is easy to quantify things. But in the order of large numbers, we lose the concept of the quantities and rather more appreciate the trends, thus qualifying what we are describing. This is related to our human perception which is what a lot of this time-space theory is revealing. We are the limited observers by centuries of paradigms of how we view and listen, measure and quantify. It is subtle, but profound in its effect upon our having the ability to understand more fully the world around us, and how we do the science of studying all there is.

For example, I measure a cup or more of coffee or tea. Thusly I can quantify it since I actually did drink three cups of such this morning. But if there are many numbers or too many to bother to count, depending upon our inclination, we may rather say, 'I had a wonderful day' (because more good things happened than bad). The exact numerations are not stated, but my feelings summarized it as a 'good day'. So it seems this is where we have the fuzzy line of self-quantification to qualification. We tend to do this in science too. Everyone has the habit of losing objectivity somewhere, sometime. We all have pet theories that we would rather prove than disprove, whilst with our mouth state that we are being 'objective' and have many other scientists in agreement with our results. I, being as human as you, tend to do the same and have to force myself to stop and recheck what I have been measuring and observing, to really make sure that I am interpreting such information objectively.

That is why I suggest repeatedly for you to check me out, try these things yourself. So often I hear others rationalize, standing only upon the colleagues that agree within a discipline. Then later through my own research, I discover that nearly 50% of such really did not agree or had queries. Also further revealing that out there in academia, another number of nearly equal amounts to the first, totally disagree. Despite the illusions of the enlightenment of democratic majorities, history more often demonstrates scientifically that the minority is more towards the human limit of truth than the arrogant and haughty majority.

Chapter Two

Non-Intrusive Time-Space Journeys Begin

It took twelve years of theoretical dissertations, seven years of testing and three years of provocative results to even get to the place to begin the primary experimentations. This all became serious study around 1998, when previously only bits and pieces of things seemed to surface. Twelve years later, profound thoughts were finally published in the first edition of *HYPERSPHERE ... A JOURNEY AT THE SPEED OF GEOMETRY* by this author. Since then we have gone farther, but still as it is written: 'We see through a glass darkly.' Even at this writing we have perplexity and realize we have so much to learn. The more we seem to learn, the less we seem to know.

Around 2010 began the juncture when Non-intrusive time-space voyaging was seriously attempted. If one wants to realize how often people actually do such, then consider looking at an old photo, listening to an old song on a vinyl record or CD. Realize that the 'now' of that recorded moment has required the energy to spin a turn-table or laser scan the CD and amplify the recorded information. The same for a video-tape or a DVD. Watching an old silent movie requires amplification, replaying energy and an effort to repeat that 'now' of long ago. Even recalling an old memory requires energy. More energy is expended to 'go back' in entertainment and in thoughts (the Non-intrusive method of time-space travel) than the original moment. Non-intrusive means that one has no kinetic/inertial interference with the prior moment. This goes for the future too, but the signal is far less in strength.

All this goes along with James Clerk Maxwell's equations based on established geometry and physics form the 18th, 19th, 20th and 21st centuries, as I have earlier eluded to. The difference is the method of measuring and observing. It is as with many things supposedly new ... they were there all the time. In order for something to be discovered in God's creation, it has to be there in the first place!

We must go back to the Hypersphere and the discs, for here one can do some fantastic experimentations. Since we have given some explanation of the use of one disc and both are the same, then that should be enough for this endeavor. But we need to delve into the Hypersphere itself and better understand this, the third of the Third Level Machines of not only geometry, but Hypergeometry. Keeping in mind in all of this, it is the first time mechanizing what is usually considered abstraction and esoteric in function. Great for blackboard or white board 'gymnastics' of fantastic possibilities; then having lunch and thinking of all as a fun time, thus remaining within our comfortable paradigms of having knowledge of the universe and well- stroked egos.

Now with such Hypergeometric, and even Geometric Non-inertial to inertial mechanism respectively, we cannot really just walk away from some of these possible consequences now in actual function of what the numbers are implying.

It is like atomic theory was on the chalkboards of the 1930s and 40s (then a luncheon and a game of cards). At the end of World War II, those numbers when mechanized, vaporized Hiroshima and Nagasaki with the loss of millions of lives. Today with Hypergeometric Mechanics more dangers have surfaced, as well as wonders. No longer with articulations of such machines will one have interesting experiences, but rather be prone to radiation damage extremes of heat and even disaster. This goes for the hobby of rocketry as well. Even playing cricket in the Commonwealth or baseball in the States can be fun or mean a serious injury. All of life is a two-edged sword. One can hide and stagnate, or dare and explore; either way, be prepared for the consequences.

I also say quite often, 'If I have been given the proper match, I shall strike it, and light it, and set my part of the world on fire!' But I warn you, I have gotten burned many times too. Oh, I have had a life full of wonderful and provocative experiences, yet I have scars. The Hypersphere is the most dangerous and wondrous of the three Third Level Machines. Be prepared, for it can bite back!

Let's go back to the Hypersphere's two small discs by reviewing the illustration on page **12**. You can see on this disc the two spherical masses connected by a spring. Through them is a small guide in opposition to the spring on the other side, where each sphere is stabilized to a non-linear motion axially with the spring. Also along the circumference are two small spheres that do not oscillate. They only inertially/kinetically interfere enough to articulate the oscillations of the two larger spheres. This has to be adjusted by the outer ring of the disc. Inside the large sphere they are set to resonance with the Hypersphere. They are parasitic, Non-linear, harmonic 'mirrors' for the driver Hypersphere.

Keep in mind that the Hypersphere itself is driven by the oscillating magnetic field on one phase of the wave being generated, whilst the other phase is driven by the local gravity field, better defined as the local Inertial Geometric. It behooves one to redefine more accurately those things so repeatedly called 'gravity' as well as the terms 'random', 'chaos', 'zero', 'infinity', 'noise' and 'turbulence'. We have passively accepted that which has been previously assigned a name to a concept, with no serious testing involved, only that which was relevant enough for the engineering level at that time in human history. Now to explore areas where time is predominate, these vague and just-get-by definitions and concepts are left for those not seeking more evidence that is repeatable and of much more difficult and higher resolution.

The last and third of these three Third Level Machines is the Hypersphere. It can be of any size: a large one was 50 meters in diameter at the magnetic driving ring, the original Hypersphere itself was 366 cm in diameter. When the Hypersphere is not driven, it sets or rests in the open, center hole of a slightly less diameter, enough to support it. When under operation, the Hypersphere will hover slightly over the opening around 30 cm or so. Inertial Geometry pulls down upon the mass of the Hypersphere and the oscillating magnetic field pushes up. Even in a non-oscillating state, the pyrolytic graphite (or better bismuth, which are both highly diamagnetic) creates an opposing magnetic field and so the Hypersphere just continues to repel the driver disc, as the driver disc repels the Hypersphere. If just ever so slightly this condition is begun, then the Hypersphere has part of its spherical surface able to protrude under the ring and therefore is very stable as a configuration. Here is the Hypersphere, oscillating from spherical to disc above the driver ring.

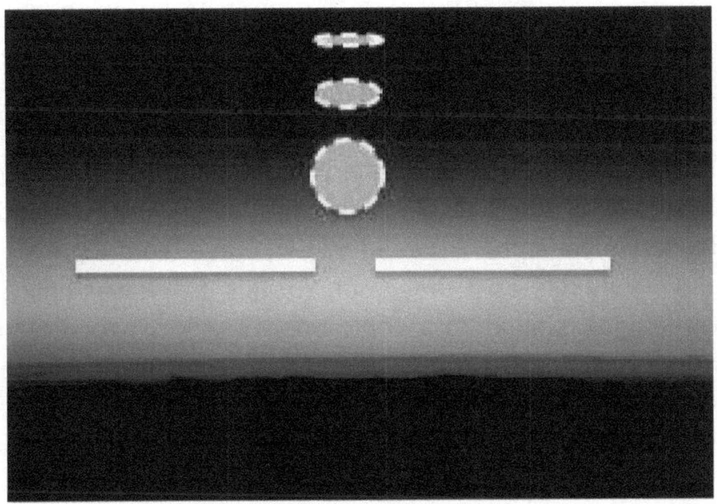

As like an antenna with a signal, the magnetic driver ring is the 'driven element' and the Hypersphere is the 'mass-signal'. The phase of push is by the magnetic driver for one half of the cycle and the pull is by the Inertial Geometric for the other half of a cycle. Oscillation is slowly, very slowly increased until Lorentzian effects begin. Then articulation commences affecting wavelength until 'slippage' occurs when the proper, relativistic extremes predominate more and more. It is the critical articulations by modulation of the magnetic field oscillations and their amplifications that aid so much. It is almost as a musical instrument with effects now appreciated. As the Hypersphere is oscillating, its mass is also undulating the local space-time curvature, which the hand-held discs easily pick up according to the Inverse Cube Law, as with tides, and can sense these out to about 500 Km away.

As the Hypersphere oscillates faster and faster, it begins to squeeze from top to bottom axially, with the horizontal diameter remaining the same. Thus mass increases, length shortens and time dilation temporally slows down, occurring within the Hypersphere. From inside the sphere, one could peer out and see the outside contracting and forming a circumference around the Hypersphere horizontally. But the outside would rather be temporally speeding up. And yet, with the outside seeming to contract, the details in the contraction would rather be very vertically stretched; while from the outside looking at the Hypersphere would be very squeezed and disc-like. As this continues, the very darkly reddened Hypersphere would slowly fade and its image would slowly rise as it faded out. For this image was only an 'after-image', similar to a sonic boom ... for it was already gone by this time! If the Hypersphere was kept at just the moment before this happened, it would be traveling into the future. Once it 'slips away' through the foliation of the local space-time, then it is heading to the past at 370,000 times the light speed and now a tachyon particle. Its rest speed is 370,000 times c. In just under 2½ hours, it is 100 year ago and 100 light years away!

I need to briefly define an orbit as 'Inertial Buoyancy'. The reason is that with Hypergeometric Mechanics, which all of this is, we need to realize that the archaic word 'gravity' is archaic. Now redefined as Inertial Geometry, we have a far better interpretation of the concepts of the local space-time curvature. With more Inertial Geometry the tighter the curvature and the lighter IG, the looser the curvature. For celestial mechanics of regular satellites this is what an 'orbit' is. It is one of many other revelations of all of these experiments and the theorizing before and after, alluding to better conceptions and definitions.

If these oscillations are kept very slow and carefully controlled, like a deep submersible gradually acclimating to increasing and decreasing extreme differences in pressure, then one will have quite the fascinating effects to witness. The danger, like the above proverbial extreme pressure submersible too rapidly traversing such changes is implosion for the submersible, and explosion for the Hypersphere. The trading of the mass of the Hypersphere to the other side of the local field of the Superuniverse's Event Horizon, is the extremely reactive release of energy equivalent to the mass of the Hypersphere. So a very small Hypersphere is not only a wonderful, delicate and intriguing instrument that is very time-predominate (actually an 'active' clock), but also a dangerous mechanism as a weapon in the wrong hands. The average home clock is a 'passive' clock. On the other hand, this 'active' clock is quite something else. Time-space articulation is as dangerous as space, the ocean, fire and flying. With proper considerations, it is a grand exploratory tool; without precaution, one may not survive or at least be somewhat scarred.

All of this becomes so obvious once one begins to delve into placing the theories under analysis and experimentation enough to actually measure and observe such. It has been so easy to say that no mass can exceed the speed of light and yet, it has been accepted that only space can. The local space-time curvature, when the Hypersphere machinery is not running, is normal and limited to the speed of light. When the machinery is under operation, the local space-time curvature affects the mass of the Hypersphere by the oscillating mass therein. Now the space-time fluctuations and its foliation are able to move above the speed of light and carry with it the Hypersphere through the local part of the Superuniverse's Event Horizon!

In this present dissertation, we will need to briefly go over the 'superuniverse' concept, and that will be done in the next chapter. The content of this book presumes you, the dear reader, has already read and understood *HYPERSPHERE ... A JOURNEY AT THE SPEED OF GEOMETRY*. It is dry theory and much more detailed. Perhaps for some it may be used now as a coffee cup coaster or a hot mat for a warm plate of biscuits from the oven. For you, the serious experimenter that you are, it is valuable to understand the Hypergeometric Mechanics that are the basis of all this.

Chapter Three

The Superuniverse and Global Event Horizon

One needs to have some idea of the 'roads' on which this type of 'sports car' goes, so to speak. The Hypersphere is a Non-linear Harmonic Oscillator stated quite simply. All we are doing is articulating it and its local space-time curvature. To arrive at this simple conclusion, we need to understand another revelation: that all background radiation may have in some cases actually been leaking radiation from one side of our local area of the Superuniverse's Event Horizon, from the Tachyon World to ours, and vice versa. This may all be far more naturally occurring than humankind has imagined.

Now thinking back upon the matter to energy conversion in the atomic and hydrogen bombs during those tests in the latter 20th century, we may consider that instead of conversion, perhaps in some cases there was transfer of mass from this side of the Event Horizon to the other; and so the result was powerful energy released in equivalence to the masses involved, of only 3% for the fission process and maybe a bit more of 5 to 7% for the fusion process. What of the inside of the stars? Perhaps higher amounts of transference. Now reconsider the black holes in the center of galaxies ... what is the transference there? Also with matter-antimatter annihilation there is rather 100% transfer as to what we may be seeing. If energy cannot be created or destroyed, perhaps matter's radiations when dense enough, are leakages. And when matter and antimatter 'annihilate', do they rather just slip through the local representation of the Global or Superuniverse's Event Horizon? Fantastic yet not so counterintuitive considerations for what scientists have previously interpreted as annihilation. We of these views are not saying that we have it all figured out, or that we fully understand all that is going on. All we are saying, is that these ideas may be the clearer view of what is actually happening.

When I related to the Temporal Diffraction Grating, I used a Newtonian Time perspective or dimension to serve all three of the time states therein, past, present and future, from an angle overhead or to the side. But actually, this is further perspective for observation. A more profound view of time that has come from our experiments is that it has three dimensions, one is of forward entropy, at 90° of geometric translation when we observe it as expansion, and at another 90° of geometric translation as rotation. Let me explain this some more.

As we are on this speed of light surface, we are also floating upon this surface of our whole local, observable universe. Likewise, others are local and observable to only those inhabitants therein. We are now viewing time spatially and thinking only temporally as the major view. So with each side of this surface having a forward entropy, it is actuality in opposite directions relative to the other. Then we observe expansion on both sides as well, and even rotation. Now looking at time this way profoundly changes any inherent paradigms of less perspective that really limits our conceptions to 'see' time as the major player in reality, whilst space is actually the lesser player.

This distinctive perspective allows different and new ideas of viewing not only motion in space, but 'motion in time'. Remember, as we easily actively travel in space, we are also traveling in time; and to actively travel in time, will mean to also travel in space. One is interrelated to the other. We just so happen to, from centuries and more of experience in humanity, far more think, perceive and act with a spatial predominance. Taking time now as the predominant brings in other considerations of very profound possibilities. This includes deep space travel as well.

Taking the dare to go beyond conventional paradigm, we now focus on time travel as the way to space travel, including deep time travel to accomplish deep space travel. As this space-time and time-space relationship protects distant places from interfering with each other, so too distant times cannot interfere with other distant times. Actually, this concurs with Hawking's Causality Protection Conjecture very well. Even nearby spaces and nearby times, though closer, do not initiate any causality problems as per the limit of the speed of light and its inverse square law of diminishing kinetic/inertial influence. In other words, the ability to influence is thus restricted with more distance. It allows there to be less of an extent to be considered irrelevant, though theoretically the signals are so weak and could possibly exist on into infinity. So the same with distant times, but starting from a closeness of nanoseconds and faster to hours and years, or more slowly and respectively nearer to farther.

Rationalizing with a time predominance does change thinker's perspective of oneself and the world and local universe around about.

Now this business of foliation and slippage should be well-thought-out. There seems at some point when the smaller the parcel and the higher the frequency, the shorter the wavelength of any electromagnetic radiations. Thus easier for those on each side of the local apparition of the Global Superuniverse's Event Horizon to leak through from one side to the other. This local Event Horizon is attainable by oscillating the Hypersphere sufficiently, and depends a lot on the local curvature of the space-time of the theater of operations (on one's kitchen table upstairs in the rear apartment where some of you may be). But wherever the local Event Horizon is, just the nearby manifestation of the

Superuniverse's Global Event Horizon or even millimeters away.

Given the provocative view that in millimeters, light years are attainable in reverse time with a rest velocity of 370,000 times the speed of light, does have quite the bewildering reaction for most people. This should account for some of the background radiation not accounted for in other ways around us. For example, in a sealed thick lead box that is also a powerful vacuum. This may all be some of the pronounced radiation leakage (so called 'virtual particles') that one will be getting from experiments with oscillations of the Hypersphere.

The mass-signal, the Hypersphere itself floating or later oscillating above the magnetic drive ring, is the payload. The magnetic ring has replaced the rocket. Essentially, the rocket is obsolete for long distance space travel at this juncture. For shorter distances in the inner solar system, it is fine; but for longer distances, the Hypersphere is better considering the return on energy investment. Likewise it is energy intensive to use it for short distances in time or space, but for the longer distances in time or space it is more economical (and the only way after waiting long enough for something to get beyond Mars). The Inertial Geometric 'gravity' field is locally adding energy on the drop per half cycle. This is quite effective if one is carefully articulating oscillations. One is actually 'modulating' the mass-signal between the magnetic and Inertial Geometric, and so it is similar to bouncing 'off the local gravity field'. Another way to understand it is as 'pulsing' the mass-signal. All this is oscillating or shaking the local space-time curvature so there is a lot going on ... 'a whole lot of shakin' goin' on!' Everything, including you, will be 'shakin' all over as well. But because the local Inertial Geometric is bringing in phase all around you, you soon are not aware of any of this shaking. Only someone at a great enough distance observing you and all around you would notice.

One of our public 'demonstrations', really a rehearsal with only those serious and willing, was quite an experience. If ever the opportunity allowed it to be done with a large number of people, I would repeat exactly what I have said to a small number of observers: "Your clocks shall slow down and so shall you; and there is nothing that you can do about it!" In this bubble of Lorentzian effects was the fore-shortening of the vertical axis, increasing of weight and also radiation increases with rising warmth. Mechanical watches were found later to have slowed down in reference to more spatially distant places. I am being more qualitative in description, for while all this is going on, it is bewildering and fascinating; you are not prone to accuracy when taking all of this in at that moment.

This is very benign compared to some of the more extreme uses for the Hypersphere. The effects are measurable out to about 500 Km, but as I have alluded before, according to the Inverse Cube law as with tidal actions.

The next level is to feel the approach of the local Event Horizon, and this is dangerous. There is a lot of radiation and the area above you seems to darken, as well under your feet, as if you were upon scaffolding. But you do not see as much if just standing on the ground, for that ground is affected too. The darkness in the sky overhead is of a distance; but if in a room, you would not see the darkening, for the ceiling above you is affected.

From a distance away, an outside observer would see you appear a bit vertically flattened and reddened, and as sort of a spherical, optical anomaly that would be scarcely detectable.

At the next level (which is ethically forbidden) if all goes too fast, then the transfer of all oscillating would slip quickly to the other side of the Event Horizon, and the energy from that side would be transferred. It would appear as a 100% expression proportional for the mass involved and your experiments and further dreams in this world would be over. So that cannot happen with all that is herein this handbook. But the potentials of such are quite possible. It would be advisable to say your prayers if you ever on your own get close to such a point as this.

I would rather you enhance what God has blessed you with, and improve the world around you. If you are not of these intentions, then may God limit you.

In extreme explorations (including Hypergeometry), science and mathematics lead to philosophy and religion. When we go beyond our reason, all we have left is faith!

Where prose fails, poetry often continues, then music, then worship! At this moment, I feel you understand me. I have the scars and burns, yet I thank God I have survived this life and so many things He has allowed me to experience. Without Him, I am nothing. You too may wind up as I, in time, I pray.

So much of this is yet as proposal; as facts, a piece here and there at times surface as theory is proved or disproved. Herein is no claim to any total understanding. It seems by the trends of things to glimpse, and that is quite the intrigue.

In time, you may prove much of this as a far different understanding of what all herein is involved. It may, in such appropriate time be a far different view than this meager hypothesis and its profound implications. So again, enjoy the ride and keep learning. It is often said that 'the more one has learned, the less one seems to know.' And it is written: 'we see though a glass darkly.' I surely concur.

Chapter Four

Where is When

Now we shall contemplate with time as the predominate, using first the Riemann's Sphere. We propose the superuniverse is closed ($\Omega > 1$) so that our little local observable universe is floating amongst some or many other universes. Upon this Riemann's Sphere of Time our local universe is the predominate 'chord' and space is the minor 'chord', dependent upon the time 'chord' far more than we have realized in human history. This temporal 'symphony' has a surface equal to the speed of light, and the direction of time on the outside is opposite to the inside. The inside is thus of a time direction in opposition to the outside. Yet, both sides need the other side to exist. Space here is one dimension upon this surface. The Riemann's Sphere is of the fifth dimension or Inertial Geometry (gravity) and anything too far above it or below it is Non-inertial Geometry, a Tachyon World. The closer to the surface the stronger the Inertial Geometric, and thus Inertia. Each side recognizes it is on the outside, and the relative Tachyon World is the mantle under the surface from each side. We feel we are on the outside, and the other side thinks they are on the outside; yet both of us sense that the underside is inside the Riemann's Sphere as we observe and measure it. World lines from past to future are from one pole to the other, depending upon which side we are on. The surface, the speed of light, is also the surface of the Global Event Horizon which appears in each local floating observable universe as the many apparitions of smaller Event Horizons (which are all of the same Global Event Horizon).

The Tachyon World's rest velocity is 370,000 times light speed and is where an object goes when it is Non-inertial (Non-inertial Geometric) and able to change world lines in the relative reverse time of either side to its respective forward time. In this, entanglement also ensues. But when an object goes into the past, then its world line automatically changes to protect causality. Entanglement occurs when an object maintains its existence on both sides of the local Event Horizon (part of the Global Event Horizon) and coexists on two world lines. This entanglement, if broken, leaves the object now one world line.

All this is quite counterintuitive at first; but bear with all of this, and what is implied in Hypergeometry and its application as Hypergeometric Mechanics will make more sense. Relativity and quantum mechanics alluded to these things, but with this detailed projection of geometry, we have something quite interesting. On either side, we can never get to one pole or the other; one is called 'zero' and the other 'infinity'. Both are beyond our reach or possible understanding. The whole Riemann's Sphere of time rotates, the second dimension of time. The first dimension of time for each side being the direction of entropy in which to each side is in opposite directions, as before stated. The third dimension of time is considered as the Riemann's Sphere in expansion, and both sides observe that. So both sides observe their own direction of time, but opposite to the other, whilst they share the same rotation and the same sense of expansion. Just ride with these repeated concepts without being able to bring them to total conclusion. All we want to be able to do is traverse time as we do space, and for this we need to focus upon near the equator of our grand Riemann's Sphere. We seem to be near or just above or below it in latitude because of the way mathematics seem to function. The red shift of the far reaches of our local universe bend over the superuniverse horizon, and the rotation may show in the bi-directionality of the Alpha Constant (and maybe other constants over far, far inter-galactic distances). Expansion seems to fit in globally, and the edge-bending apparition of red shift is near every local universe's observable edge, until light is of too low a

Doppler shift of frequency. We do not know all the 'whys'; we just try to make some sense of the observations and measurements so far obtained in the past century and this one. Illustrated next is the Riemann's Sphere graphing temporally 'where when is'.

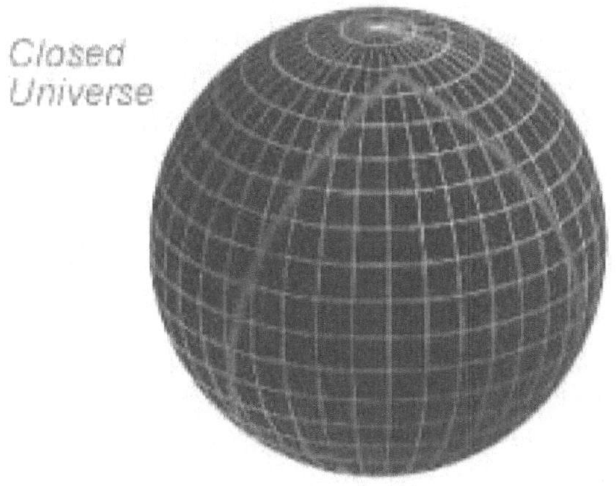

The 'Closed Universe' may prove to be subsequently true, but we are supposing at this writing that it is more a 'superuniverse'. Perhaps the local and observable universe that we do know is more like this after all.

These concepts, still so difficult to comprehend, seem to apply to our projection of geometry to Hypergeometry and its being as mechanism to 'active' clocking of a Hypersphere, a Third Level Machine. Therefore this may be the map of time in three dimensions, space reduced to one dimension upon the surface and the fifth dimension as Inertial Geometry where inertia and kinetics exist. Keeping to our outside orientation, (since inside the sphere is the tachyon mantle to our side and above it is decreasing inertia above the surface) when we approach the speed of light to the surface, mass increases, length foreshortens and time dilates. Speeds above the surface on our side are decreasing from light. So we see a different perspective to motion, velocity, time and space in an advanced geometric sense.

The red lines are the expanded triangle of greater sums than 180° agreeing with this positive and closed curvature over superuniverse distances of space. Also it helps to illustrate the changing of one world line for another when any object goes back in time and begins to go forward for that same side, which we are defining for ease of thought, again as 'our side' versus the 'outside'. Then, at 370,000 times the speed of light, in reverse time for the object considered, it is only around six minutes to Alpha Centauri. To go back 100 years ago or 100 years into the future, will take nearly 2½ hours for this object to traverse within its own internal time passage reckoning. This includes the occupants of such an object. Even you!

With the Hypersphere, even macroscopic objects such as you, are now able to do what is happening at very relativistic velocities and in the quantum realm of things. The concept of 'when' can be specialized for our very limited human thinking, and this Riemann's Sphere may help us to make more sense of things. Thus, the question of 'Where is when?' is at least somewhat clearer upon this projection of geometry to Hypergeometry.

By the way, I offer entanglement to possibly be equal to tunneling. For in the entangled state the object is in two places and two times at the same place and at the same time. The corollary to the two places at the same time is two times at the someplace. Normally as you sit in your chair reading this for so many minutes, you are at different times at the same place. In the quantum world and extreme relativistic work you could also be in more than one place at the same time. Tunneling, we also conjecture, is happening at 370,000 times c and in reverse time, even in tunnel diodes commonly used in microwave and radar applications. Tunneling and entanglement may be the same creature, but experienced from a different perspective, all wave-particle duality.

Also, I have kept referring to our local observable universe and others as though isolated floating upon this surface. Perhaps, it is the whole universe with our part still just a fragment of this universe which we are limited to observe and measure; so we kept the illustration's title above as 'Closed Universe'. After more experimentations to come, we may rather consider this as the more valid. So then superuniverse may really be our one big 'Closed Universe', so making it far more profound than we yet aspire. Lots to consider at this writing not only for you the reader, but still for me the writer!

As Al Jolsen said, "You ain't seen nothin' yet!"

Chapter Five

Some of the Shapes
of Things to Come

HG Well's wrote *THE SHAPE OF THINGS TO COME* in 1933, and though primarily a future history focused on socio-political interests, I often wondered of the technology that would be revealed from the time of my childhood into the future. Well's title also evokes for me the thought that these Third Level Machines herein are possibly some of those 'other shapes of those things to come' in the realms of more advanced methods of propulsion ... especially deep space propulsion. Time and space being so intertwined easily allows time's travel considerations when physical conditions are very extreme.

Now we come to application. I have often said that 'The application of theoretical geometry to practical mechanism is provocative.' And it is when geometry is forced to stretch to Hypergeometry by local manipulations macroscopically with appropriate use of machines meant to do such as that. That is what the Hypersphere attempts to do. The lessons learned from the Temporal Diffraction Grating and the Hyperplane thusly so led to the Hypersphere. The Hypersphere along with the discs that accompany it, are the final level attained for such dynamic mechanizations. The discs need the Hypersphere through a strong enough resonance, though they can surely resonate from another proper tuning to any other oscillating mass. It is the appropriate modulation and strength of the Hypersphere's effects upon the local space-time curvature that are so dramatic.

You, the hobbyist and reader are the player of these new geometric to Hypergeometric instruments. The 'symphony' aforementioned, is but one of many within the space-time continuum. Rather, with time predominance, it is then better understood as the time-space continuum. The relevance of the Riemann's Sphere as per function thus outweighs the paradigm of spatial predominance for the centuries before our development of these Hypergeometric mechanisms.

To empower our beginning forays into the local time-space continuum for your kitchen table laboratory, you will need some electrical power in the form of an oscillator and an amplifier, and magnetic amplifier. Also remember the local Inertial Geometric will provide the other ½ cycle per oscillation. It will be fascinating for the first and the course of other experiments that there is simplicity of structure and function for macroscopic affectations, whilst for so long humankind has missed such.

One does need billions of Euros, pounds or dollars, but does not need large bureaucratically burdened administrations, nor do you require to safeguard your career for tenure. You can already begin to experience quite quickly, without a vote from the House of Commons or Congress, your own small level of Hypergeometry in the extremes of geometry at our normal scale of existence.

Past, present and future are more open to you, more than to the present-day elites of our modern empire's various propaganda ministries of the East and West. Future history had been offered fictionally in *THE SHAPE OF THINGS TO COME*, and actually had some of reality, just as past history has been and as the present flow of 'nows', the moment to moment unfolding of history been revealed. All this according to Einstein's Theory of Block Time. Biblical Prophecy so thoroughly studied and supported, also shares the determinism of the things to come, as those things that have passed. Consider Lloyd's work at MIT, along with some colleagues on Post Selection in logic and quantum mechanics. This all herein seems to agree quite well with such a conjecture too. For this seems to imply a deterministic implication if one extrapolates to the sequence of events around us. Such considerations from the microscopic to macroscopic to cosmic scales as inherent, especially contemplating extreme physics of relativistic velocities near absolute zero cryogenics and theorizing the very small upon the very large, as past and future get blurred.

I have mentioned the concept of determinism which for many is distasteful. From these studies of Hypergeometric Mechanics, the mechanization of Hypergeometry, we view our limited 'free will' as minor to the major of deterministic flow from past to future in our local universe and the Global Superuniverse. We are accountable for those things which we have the ability to decide upon, proportional to the closeness of inertial and kinetic consequences of our choices.

If you are able to on your own disprove such as just proposed by this author, please try. Perhaps these technologies and concepts, as a result of the serious search through these works, have far more relevance than many shall dare to accept. Then if one can with repeatable, verifiable, scientific evidence dispute any herein, please do. It may be easier for many to continue to deny that which can be proven, resulting in each of us having our own 'religion'. Here religion is defined as a 'system of ideas'. To deny that which is proven scientifically and with true history, is for one to choose to be superstitious. For many, denial is their temporary medicine of choice. Your choices seem to reflect those amongst world lines and so the flow of history you experience is very consequential to your decision. It is written: 'one reaps what they sow.' Cause and effect is reality … from moral, political, social and scientific experience. The same for little ol' you. Your brain seems to be a quantum machine which decides the world line of experiential choice amongst the many nearby similar, but in time divergent other world lines.

From sources such as HG Well's volumes to the *BIBLE*, with the books of Genesis to Nahum to Revelations, think tanks of today are trying to sense trends from past to future history of events so as to profit from them. Humanity treading upon these waters of what is here considered as Hypergeometric Mechanics, shall find many of their pet paradigms profoundly challenged. 'Welcome to the future, and the past, and the present', may I firmly offer.

Whence to have speculated upon my father's proposal to me as a child (can you think of a machine with only one moving part?), I look back now and think of these Third Level Hypergeometric Machines as really each only having one essential moving part. In the Temporal Diffraction Grating, it is just the laser beam, the simplest of the three. In the Hyperplane, it is the rotating mechanism of the laser beam, so mechanically just one moving part. And the Hypersphere has the oscillating mass essentially as the only moving part. I may be stretching definition some and maybe too much, but at least it begs the possible consideration of such mechanism.

It is amazing to note not much of the Hypergeometric mechanism can compare to the now typical machinery used in the last couple of centuries.

Since we have observed that mechanism extrapolates in arts and function over time, then a new machine with only its necessary primary parts becomes more complicated with subtle adjustments and tasks. The complication rises though the basic tenet is still within its earliest structure and function. Even life seems to demonstrate such with simpler life leading to more complicated life in structure and function.

We perceive gears, shafts, pulleys, simple capacitors and diodes, then incrementing to motors, automobiles, radios and computers. Machines growing similarly to living things.

The simpler always outnumbers the more complex, as the more complex is composed of the many simpler, as a trend.

Trying to better explain things, keep in mind that entanglement and tunneling are at the same speed as the tachyon rest mass of 370,000 times the speed of light in reverse time. Along with this is the super positioning occurring at the same time, but not just in space, but also in time much more than the normally experienced in our world. How is that to be understood? Well in our normal and 'decoherent' world, we can sit in only one chair at a time, but occupy or superposition in time a lot. We can sit in our single chair for an hour, which is sixty minutes of time at one minute per minute. Also remember that our oscillating Hypersphere is also acting as an 'Inertial Mirror' where the magnetic pulse opposes the Inertial Geometric Curvature pull, similar as a mass-laser. Again, think of tunneling and entanglement as more of the same thing, than different.

The Inertial Mirror concept is similar to the Inertial Buoyancy which equivocates to what we think of as an orbit, such as a satellite going around the Earth. Inertia, from the perspective of it being considered as an active geometry, appears as the familiar, initial resistance to being moved or the familiar resistance to being stopped or 'falling' in a gravitational field. These older definitions seem to represent what we don't understand, 'mysterious'. Whilst with a mathematical basis for definition it is more functional and hopefully clearer to what is going on.

Back to 'decoherence', it is the realm of the macroscopic world in our local familiar universe. But near absolute zero, within the extreme velocities of relativistic physics, we seem to see more and more of coherence. It is as though the Quantum world of the very small is now becoming apparent at our scale of experience.

In coherence, matter can occupy the same space at the same time and experience entanglement and tunneling. This repetition of mine is to help elucidate what the Hypersphere is doing of itself and the local space-time around it.

With a single disc or both discs, one can now make use of the effects of such at some distance from the Hypersphere with a camera, audio recorder and even oneself, if you dare! This is because the local space-time is fluctuating. When it is oscillating at a certain frequency, depending upon the mass densities locally and the amount of curvature of the Inertial Geometric, then anomalies can be of such affectations. With this leakage, radiation increases as well. What is usually not our normal experiences with our day by day existence, now become more profound, and this, exponentially.

In seconds the reality of past and future and the world of extreme geometry begins to seemingly surprise one's senses. Resonance is experienced by the various objects, starting nearby and most strongly, and then with some dissipation with distance. Objects will shake, seem to fall, then hover, then continue to fall, and winds will stir even within confined spaces. Sometimes if the changes are more sudden, one is thrown to the floor and furniture shifts as windows break! The house can shake and when the Hypersphere's payload (its sphere apparatus) rises and soon flattens and reddens, there is a point where all inertial/kinetic turbulence suddenly stops. As the sphere then looks flat like a disc and is deeply reddened, it appears to rise and fade, but it is only the after image. With such an apparition, it is already moments ago, gone.

This is now the new state of affairs, for since its departure, about every 1½ minutes one light year is traversed in space with around 1½ minutes in reverse time! Thus Alpha Centauri is merely around six minutes away, and six minutes ago! All these are finely dependent upon the local Inertial Geometric curvature, which in a fine sense seems to vary as waves upon an Inertial Geometric 'ocean'. The exact figures are difficult to assess because so much is contingent upon these ever changing conditions of the local space-time.

When you get to the point of even initial stages of such experimentations be prepared to struggle to get more refined quantifications. The 'ball park' and the 'seat of your pants' seems adequate, though not so academic for finesse, for such measurements.

The resonance aforementioned, can be especially appreciated with one or both discs, where they can hover quite interestingly and seem to 'ride the local wave' for a bit of time. At the point where the sphere goes to disc and just about as it rises to drop, the disc will let it hover a moment, then it will continue to the floor or table top.

Heat is associated with this too as is ultraviolet light, so one may actually tan some. All of this should be undertaken in a remote area and preferably by one not planning to bear children. All effects are temporary, but briefly intense, with the radiations. But with personal and recording devices the interesting becomes pronounced.

Embarking

Prior to 2010, I had put feelers out to potential interested parties of academics and governments, particularly space programs, even military, but had few returns of interest. My research was still in its infant stages, but theoretical and experimental information was offered. It was quite extraordinary for sure and understandably would need more explanation. But within the United States with all of the budget cuts, the lack of interest in the exploration of space, beyond those who already were the established inner circle where the system was well operating, I heard almost nil to encourage me. I went ahead and copyrighted all of this as Hypergeometric Mechanics.

Later in 2010, I had decided to send the most pertinent and still unpublicized to Dr. Jacob Beckenstein at the Hebrew University and Israel. If there is a nation with the most promise of such, it would Israel.

After this, I went on to publish my first book, *HYPERSPHERE … A JOURNEY AT THE SPEED OF GEOMETRY* (original and revised versions). The revised is available online and the original is sold at lecture until enough are left for my private library.

The original is also archived at Columbia University's Mathematics Building in New York City, NY, The United States Naval Institute of Annapolis, MD and various other universities here in the United States. Many of these books I delivered personally to these institutions over the last few years.

All of my publications are available at **amazon.com** as well as with Kindle and in most quality bookstores world-wide.

We have a website **www.cronkhitetechnion.com** or just search for 'Cronkhite-Technion'. We try to keep the site updated well with text and videos. On my own, I continue theory, experiment and prototype work.

With all this, I ask you the reader to please build these things, experiment and gain understanding, despite the disdain and ignorance of the world to your work.

It was the Swiss father and son, Auguste and Jacques Piccard that developed the Bathyscaphe. Later, the United States Navy bought it and implied it was American, whilst it was used to explore the Challenger deep in the Marianna's Trench in 1961. It took European innovation to contemplate and develop for serious endeavor such a craft, as it was for rocketry. When one of the 'world powers' get ahold of something that opens up new frontiers, it seems that they claim it as their own. Sadly, oft times it was the lone thinker, writer and/or inventor that really had to 'courageously go where no one had gone before' ... before it is usurped by a government's military or commercial profiteers. Don't be discouraged though; think, write and build. Learn your lessons from as objective history as you can. There were those before and to come after you, that God has given resolve and talent and fortitude to withstand others that shall try to discourage.

I referred in chapter one of this book to an article that I had written for the November, 2014 issue of *Spaceflight*, a magazine of The British Interplanetary Society of which I am a member. It was high-lighted on the cover as 'Tachyon Travel'. So please check out this interesting view of some of the implications for the underlying world that seems to be the mantle of the superuniverse, upon which surface our local universe is floating.

What I find so exciting is that in Great Britain one finds the ease of conversation and sharing of what in the United

States is considered of little or no interest. I am also a member of the British group, The Wellsian Society and here one can converse and share upon the works of H G Wells. *THE TIME MACHINE* and *THE SHAPE OF THING TO COME* were such an influence upon me as a child, and along with Albert Einstein's *RELATIVITY,THE GENERAL AND SPECIAL THEORIES* made possible the serious considerations that in time led to Hypergeometric Mechanics, and the mechanization of 'extreme geometry' as Hypergeometric Machines.

I share all this in order to empathize with you, the reader, as a kindred thinker. It takes daring in the face of rejection to pursue who you are and your gifts. In time, by the grace of God, it will be your time to be heard and seriously accepted.

So many individuals and groups think that writers, inventors, scientists or thinkers are odd people with no feelings ... that we are cold impersonal sages who have 'not got the time of day'. We need to address the fact that our emotions and actions are fired by those passions that are quite eccentric, yet profound.

Nope, we're not done yet!

Chapter Six

A Daring Lecture
at 370,000 Times
the Speed of Light

Perhaps in a field in Ireland or England they have dared one of our kind to lecture. You and I are not typical. For me, if God has given me the proper match, I shall strike it and light it and set my part of the world on fire. So you hired me for lecture ... then I shall deliver. Please permit me a bit of poetry ... for this dear world, that is almost all that it has left.

Perchance it is in the British Isles or in the countryside not far from a farm. There are many coming to see for the first time what countless have not understood. It is a public demonstration before the constable or the police arrive. Many have come from great distances, even from overseas. It is geometry, but at the extremes, thus Hypergeometry. Heat of nearly 100 million degrees will be attained in less than a thousandth of a second. If something were to go wrong ... but I have been hired to lecture, then so I shall.

When the opening of the lecture is concluded the Hypersphere is initiated. Cryogenics bring extreme cold to the magnetic drive disc. It will take a few minutes as I continue the lecture. Biblical and Shakespearian quotes pop out, as the prologue is presented, for there is not much easy to say. Many are not sure what is totally going on ... it is the beginning of new worlds. Even I am not qualified to understand all that is going on. I merely demonstrate what happens at the approach and sudden achievement of 370,000 times the speed of light. It is now beyond the scientific theory and the mathematics. If the authorities are worried about catastrophe, what are they going to do? Outlaw mathematics?

The Hyperspshere is to bring the close approach of an Event Horizon. Many have begged for proof, but they never really knew what they were asking. Well, for the confidence of my clients, my audience, I shall deliver. They have paid for this, and I shall not disappoint them.

I never was popular, so I have not dodged the qualifier of being controversial. I am not here to be accepted. I am here to be provocative.

 Yes, they have asked me to prove things and so I shall. And if something goes wrong, somewhere at 100 million degrees C and in less than a thousandth of a second, I will be the first to go … and then you. Dare me to go on!

Your first alarm will be when nearby things start to shake and rattle, and the heat starts to increase. You think you are civilized. Let's see how civilized you are when some of you panic. Fascinating, when one is challenged to prove something, then when it comes down to personal danger, they run.

Now the Hypersphere is oscillating and reddening. Slowly, it will go from sphere to disc as it rises and the radiation increases. The audience does not understand, for they complain. The heat rises, they want to see the future, but they have hired me to light a match that now I cannot put out.

The winds start and all is shaking. The winds are circular around the Hypersphere's circumference, swirling and vastly increasing with strength. Suddenly papers, bits of debris, dust and smaller wind borne objects whirl within a radius that drops in influence as distance is extended from the Hypersphere now spinning and axially shrinking, with constant diameter only in the horizontal. Now I drop the demonstration disc from my outstretched hand, it hovers, then soon drops as the Hypersphere goes through its phases, from sphere to disc increasingly reddening, slowly rising up. Now the disc I had in my hand finally hits the floor, then the flash! The Event Horizon is close as the Hypersphere leaves, its afterimage fades as it is already gone! Everyone runs from the heat and prickle of radiation. Some who had said they would dare the effects will not now be able to bear children. They wanted proof, and now they have gotten it. This is the less brutal of demonstrations. The most benign was to slow clocks and themselves for a radius of 500 Km, but that was not enough. Now they wanted and got much more.

As every 1½ minutes go by about one light year is obtained by the Hypersphere that has just left, but in reverse time. The audience is confounded by what they begged for. They have so much to process and reflect upon, but they have met the future and the past and the present, presented mechanistically before them. Most of the spectators has not understood much at all.

So where is tomorrow, and when was yesterday? You have finally opened a package from God that you do not comprehend. And you thought flying to the Moon was rough! You arrogant and pious humanity. You mock the wanderer, the misfit and those of whom it is written, '... old men shall dream dreams, and young men shall see visions.' Yet you yourselves cannot fathom what you have ignored, and now suddenly you want to have all knowledge.

Please accept my drama, but it is when and where emotion overwhelms reason that the excitement of experimentation and exploration commence.

As if a transparent apparition, the disc's red column to after image, to fade then gone, has left a slight disappearing trail. It has already been to before it has left! Now all is quiet, but disheveled around the humming driver disc, but the launch area remains still a somber place of restfulness. No longer restricted by lack of perception, in reverse time, but at 370,000 times c, the Hypersphere's course directly above its place of dormancy, onto new times and places, supposedly 'unattainable'.

Who am I to be an agent for such exemplary experience? For having lost everyone important to me in the last several years, it was now for me to gain so grand vistas. No longer hindered by doubt, cowardice and great loss, renewed courage incited me to participate with indulgence ... time and space; the past, present and future, and other places light years from human history and travel.

What privilege is offered unto myself and others as misfits in popular society? Abandoned to fend for our dignity and value, and for great expanses of time to second guess ourselves unto repeated defeats. Now, left to ourselves, to build these machines that are not seeming to be understood by most others. Even governments and universities need time to render what worth geometry as mechanism really means.

Also, in my repeated reference to our 'superuniverse', I perhaps should just be saying 'Universe'; for everything about the Universe is far beyond us. We are finite and the Universe is infinite in relation to our interpretation, based upon our limited understanding of an infinite superuniverse. We stumble with definitions and conceptions because we are immensely more limited than we choose to arrogantly believe. We only glimpse of greater things, despite our haughty attitudes and self-celebratory indulgences. Instead of so easily accepting our trophies of grand endeavors, we should rather sit down for some moments, reflecting upon truer versions of humanity we find distastefully humbling.

In all of these things may we see and hear and understand as it has been said, with eye and ears and minds that constrain not our comprehending. We often are limited by our affections, or better, limited affections. We sometimes stop in our quest for knowledge and academic achievement, choosing to rest along the side of the great road to wisdom, merely satisfied with our discoveries and accompanying accolades. Temporarily it feels good, testing intrigues whilst to only truncate the fuller vista we now are blind to see and deaf to hear.

Chapter Seven

Is Future Past/Is Past Future

It is with great significance to render thoughtful resolution upon all the refined results of the measurements and subsequent observations. It is upon the processing of such glimpses of time and space, along with their implications, to better understand our place of when and where in the now and here, our 'present'. Then we are able to put into context our more clear 'present' in the greater scales of time and space.

Think upon this: We can easily consider of history and the past, and even of the near future times to come. Now perceive from above such locality of time and space and think anew. If I reflect upon what I might do and say in a hundred years, and really kinetically interfere, it will be for the historian and in a thousand years for the archaeologist; then in a million years for the anthropologist and for the paleontologist with increasing factors of tens, hundreds to billions of years for such. Now look backward in time and we see the mirror of this. We do this with space too. We are so local and yet we think about traveling several kilometers for business or for holiday hundreds and thousands of such units. Now to go millions and billions, we are no longer the common practical, but would consider space exploration. Therein are our conceptions of time and of space, compartmentalizing local and far away. Locality is our everyday reality, but so is the 'far away', whilst such long ago of time and space seems so irrelevant experientially.

All of this diatribe is but to emphasize how we too easily perceive in convenience. And this is fine for most practical purposes, but to glimpse the grander scales of time and space we need to understand what the so much greater resolutions are revealing to us.

The 'silence between the silences, and the shadow between the shadows' implies that more is to be measured and observed than has been historically respected. It is when one could say art and poetry are seen and heard, even inspired; heard by many, understood by so few. This so quickly blinds and deafens us from such subtler things that for all practical purposes seem sufficient. But to dare and go further takes far more tedious, greater, resolute measurement and observation of experiment that draws one away from more easily achieved successes.

One is a 'stranger in a strange land', as it is written. This experience of a predominance in a perspective based on time (space-time or now time-space) vastly opens one's eyes, ears and mind to that which has been so subtle. It is now with a far more Hypergeometric perspective that one has not ever really taken seriously enough. For those of us who bother to think beyond the envelope, especially inspired by science fiction in our youth, can more easily accept this concept. We of humankind, so limited by the everyday, practical things that demand our attentions in the very silent dark of the sound and flight about us, begin to perceive ever so slightly the substrate of the mathematics basing the fabric of time-space, then space-time, then matter and energy.

Gravitation, so often difficult to explain, whilst its common manifestations seem easy to experience, has needed redefinition. It has begged our deeper contemplations and analysis, yet all too often we find ourselves back to befuddlement and then just refrain from trying any more to understand it.

Here it is quagmire because of our limited definition of it. Replaced by Inertial Geometry, which does need the repetition in this work, for slowly we shall see and hear and glimpse far more than we have dared to before of this new 'land'. The substrates are the Tachyon World of time reversed particles, because they exist in a time reversed mantle of the superuniverse, then we have the speed of light surface and our familiar world of forward time traveling particles. All this has an Inertial Geometric method of structure and function.

In other words, let us go back to a reality first created as mathematics, Hypergeometric Geometry, where Non-Euclidian geometry is the norm. Then we have islands of 'gravitation', really Inertial Geometry. These seeds are the seeds of the many island universes floating upon the speed of light surface of the superuniverse.

There is a first basic mathematics, then time-space, then islands of space-time, then matter and energy within each island universe. Ours we call the local observable universe. All these island universes have forward going time and at the speed of light surface, no time, and below the speed of light surface, reversed time.

Now let us pursue 'gravitation' with a better name having a structure and function called, Inertial Geometry. This would be the fifth dimension if you will to the other four dimensions of three for space and one for time. All dimensions are reversible, even Interital Geometry. There is a Negative Inertial Geometry which may help the Inertial Geometry of the rotating superuniverse maintain itself. The superuniverse is a rotating black hole with a speed of light surface. All black holes are the same thing, and in the same place, and at the same time. Though upon the surface of the Super-Black Hole Universe our common experiences of our local, floating, observable universe is the same as the others.

All of this provides the stage for our very small act of attempting to traverse within 100 to 200 lights years, and thus 100 to 200 years of forward and reverse time travel, with the required correlation of space travel. In other words, for me to trace forward or back 100 years of time, I need a space of 100 light years. Remember, that the Hypergeometric velocity appears to us, yes appears is the word, for we have only our reference of the now and here from which we have embarked. The Hypergeometric Velocity is at 370,000 times the speed of light. At this velocity, forward time and reverse time is achieved at slightly close to c for a material object, with an abrupt pulse in synch with the oscillation occurring at the prior frequency for the object's mass, and requiring the appropriate DeBroglie's wavelength.

I apologize again for so much repetition in the writing, but it is essential to actually understand and accept what at first is very counterintuitive. It is as essential to understand celestial mechanics if one is traveling through space, compared to traveling on a bicycle upon a country road to grandmother's house for lunch. To ride the bike is comfortable, familiar and understandable as to require very little change from walking the same. But to travel to other planets in our solar system, as our technology has enabled, is quite counterintuitive to everyday experience as to tax those who are still bewildered by spaceflight, even after the mid-twentieth century with its many accomplishments.

Imagine trying to relate all this to a person of the latter nineteenth century. At that time technology in the western world was beginning to touch upon such celestial mechanics concept among many of the educated. For some, they could grasp it all. Now include H G Well's great book, *THE TIME MACHINE*, and we have another level of the counterintuitive to try to understand, and even a lesser number of people fathoming such. So many, even in this twenty-first century, are not grasping much more than a child's appreciation of the most popular space travel achievements, and with a lack of concept of the realities of such abilities within our own solar system.

I feel that if you are reading this with a scientific and mathematical hunger, you also realize that many around you are aghast at those unguarded moments when you slip your guard and try to share some of these grand innovations in the current events of the day. You have realized that often you wished you had kept silent, but in your excitement you could not help but share such wonders to family and friends, only to leave them staring with mouth agape, silent for a long moment. Suddenly, they would choose to chat upon something far more mundane, leaving you to reflect on your 'faux pas'.

This is also the bane of many of us who are attending university or in corporate research. Administrations in such require balanced budgets, much from the national military-industrial complex per country of your personal abode. This takes precedence over our personal attempts at funding our experiments in areas which are considered 'on the edge'.

In this work, I am trying to allow you on your own with limited funds to actually test and try to prove me wrong or right, as you choose, these Hypergeometric Mechanics.

Pardon my tangent, but I feel you as well struggle in all these similar areas, since your own beginning of personal studies in theses extremes of physics and mathematics. There is so much implication in what we seem to be glimpsing that it will often require you to digest, possibly with a glass of your favorite wine, what you are slowly measuring and observing.

In quantum mechanics there is non-locality well considered. Also we often hear in such that at extreme cold, near absolute zero as in Bose-Einstein Condensate, objects occupy the same space. This is a spatial coherence which we in the decoherent macro world easily see. I am sitting in a chair for fifteen minutes. Each minute while I occupy the chair, my space is in more than one time. So my one space has occupied more than one time. Now there is a corollary to this that I can occupy more than one space at the same time, if I approach absolute zero in my space.

Thus, an object can occupy more than one time in one space; and an object can also occupy more than one space at one time. Actually we are limited to only two places; that is where entanglement comes in. If we take our full human body and consider we are each cell (again definition is important), then I guess we can say we are in more than one place at the same time. This is far more than just carnival humor; it is very interesting.

I can place each of my hands on two different places and be in two places at the same time. Such is required for my ability to ride my proverbial bicycle to grandmother's house for lunch. Everyone takes all that for granted, and so as a paradigm, one never contemplates such a situation more profoundly. Here definition and scale are important in our perspective to better glimpse the subtle effects of Hypergeometry and its appearance in our decoherent world at the macroscopic scale.

So even at our scale of reality, there are shadows of the quantum world and the relativistic world. Our common world of forward time is also shadowed by reverse time, and the profound lack of being able to know which way time is flowing.

Look down upon a billiard table without pockets and film the balls bouncing around. Put the film away for a month and then review it, but have someone else run the film, choosing to run it forward or backward. You will not be able to tell which is which. For without a temporal reference to a sequence of events, time at our macroscopic and decoherent scale is also forward and backward in some observations and measurements.

Time and space, and even gravity have forward and reverse directions. Bi-directionality is fundamental to our universe and superuniverse under some conditions.

When we oscillate the Hypersphere, all we are doing is placing it under an extreme enough state as to allow it to travel the easiest path, the path of least resistance, in time and space.

There is a lot of Inertial Mirroring going on in our time-space. If we oscillate our Hypersphere to the extreme at near light speed, the attainment of 370,000 times the speed of light occurs for forward time, for it is a Lorentzian effect. When we cross the Event Horizon we are suddenly at 370,000 times c and in reverse time. This border-land of 370,000 times the speed of light and the imperceptible crossing are both near light speed; we travel just near the speed of light. It is after we hit some grey time-space that we are suddenly at 370,000 times c and for some moment in forward time, and then so very suddenly in reverse time. We are vastly traveling in references to our surroundings at this speed, yet very coherent and non-local in time and space. We are experiencing entanglement. This entanglement is a realm when forward and reverse time are occurring very close, and we choose to 'bump' by modulation to the area that is reverse in time.

This 'bump' in modulation is needed to add just enough energy to the oscillation to move it into a coherent state. Modulation happens as other frequencies are amplified during oscillation in the Terahertz (Thz) range. The frequencies are from the high microwave to Thz to lower infrared ranges. Each object, according to DeBroglie, has mass and dimensions of particulars that are in axial motion of the oscillation's main frequent that 'fits' the particulars of the Hypersphere. Producing a rotating magnetic field can also be used and so one can see that the 'bump' is actually an extra sudden burst of added energy that carries the Hypersphere over to a coherent state suddenly.

In 2010, with very little interest returned from our intense canvassing upon those in the United States, Great Britain and others of our allies, only Israel was finally interested enough to receive the first drafts of Hypergeometric Mechanics. Essentially, they are the most mathematically powerful nation on Earth.

Consider this: that though at first outside interest was scant for all the concepts (though based on established mathematics and science since the 18th century), it was and still is very new and quite counterintutitve. Yes, it is deeply profound and yet so attainable at simple levels by any amateur that wants to test these things. Whether governments or individuals, anything new is bewildering and sometimes even upsetting.

Your experiments with your own Hypersphere are to be your individual explorations into these extraordinary worlds and you will become another 'stranger in a strange land'. It offers a very broad field to attempt your own discoveries. I admit to having just glimpsed and to have understood a paltry amount of the greater picture just viewable. Take your own attempts personally and let no one dissuade your efforts.

Let us in some moments try to seriously now go over your own Hypersphere with discs that you will build and operate. It is even advisable to form a team so that each of you can make the efforts more bearable than for one 'lone wolf'. If the solo method is fine for you, that is wonderful. Do what works for you the best.

The details are needed now to actually attempt all that has come before in this writing. You will need our aforementioned Hypersphere book. A smaller version that can set upon your kitchen table is a great start for your

Hypersphere. Even one with a driver disc of only three tenths of a meter is fine. If your hand held disc is nearby, you shall easily be amazed at the physical consequences within your 'laboratory'. Effects should be very pronounced once all is underway.

We do need to realize that our own perceptions of past and future will be quite changed. One will easily begin to realize that we are all time-space travelers, and on the most subtle levels much of this has been there all the time that time and space have been time and space! One of the most jarring of concepts to accept is that matter is quite much more space. Around 99% actually. So consider this: that under some condition of extreme geometry you may find that objects are observed passing through one another. And confusion will result if something has happened before or after something else in a series of events.

Under oscillation some objects will appear in two places at the same time. It is the effect of entanglement and the nearest thing to such is the so called 'quantum ring' at the nano-scale that has been observed in the last several years in quantum mechanical devices in atomic scale physics. Take an elliptical ring of atoms and make one atom occupy one focus of the ellipse, suddenly another atom just like it is observed at the other focus. It is at such a high frequency that it appears as one atom occupying two places at the same time. In our articulations for macroscopic objects we will be dealing with far greater scales and trying to conceptualize at even light years.

The only interference kinetically is at the two foci of the line of entanglement. Not only in space, but in time. Why entanglement seems so 'instantaneous' is because it is happening at 370,000 times the speed of light and in reverse and forward time at such an oscillation over such a distance. This begs the question of a wormhole and tuning of this object now occupying two different yet alternating places, at two different yet alternating times. In the Thz range this is happening very fast.

Around 1.5 minutes per light year, between two points in space and two points in time, is a very good gauge of profound effect. Each point is either here or one light year away; either now or a year ago, or now and a year to come. Alpha Centauri is around six minutes away for around 4.3 light years; 4.3 years ago, now, and 4.3 years to come. All depends upon the points of reference to and from where one is measuring.

This is what a Hypersphere does. It uses nonlinear, harmonic oscillations to affect quantum and relativistic effects in extreme space-time geometries, which now are time-space, because of the predominance of time over space. In our normal everyday world it is space predominate over time.

There are more illustrations to come and perhaps these shall better make some sense out of what all this implies of our local volume of our observable universe and very long distant space travel by non-linear means. Up until now, most space travel is by linear kinetics and limited thus to the speed of light, and having the increasing struggle to try to approach the speed of light with increasing needs for exponential energy requirements.

It is in our local volume of the universe out to around 100 light years or so, with 100 years past and 100 years future near the benchmarks for our own experiments in Hypergeometric Mechanics. It gave us wonder and some semblance of familiarity with family history, local personal interests, and keeping things within ranges of power and application that did not require prohibitive energy.

Going too far in space and time, and too small or too large in scale, severely confuses past and future and spatial familiarity. One is dealing with tunneling and entanglement and this is bewildering to some extent for anyone. We are like children playing with fire; it is nice and warm, but it can burn too.

Picture two spheres, both light bulbs really, and not too bright as to blind you. Separate these two spheres apart about one meter. Now align them in front of you upon a table in this thought experiment. Look just above them and if you want to the side of them slightly off center to the alignment. I would like to show you what is quite subtle, yet profound. In front of you is a simple demonstration of time and space travel.

With each three tenths of a meter, we have a photon traveling at the speed of light, and it covers this distance in about one nanosecond. So what you have before you is actually a clock of sorts. It is also a view of past and present and future all at once. You are having a perspective just enough to glimpse such. How is this to be?

Please let me explain. Let us call the farther away bulb 'far' and the other closer bulb 'near'. Sounds too easy and almost silly right off, but bear with me. When you look at the light coming from the 'far' bulb, it is coming from a greater distance limited by the speed of light, and so is farther in the past. The light from the 'near' bulb is less of a distance away and therefore not so far in the past, all to your reference point. The light thus from the 'near' bulb is in the future to the 'far' bulb. You are seeing light from two different times in your past. Also the light emitted from the 'near' bulb is in the future of the 'far' bulb, for it is closer than the 'far' one. Remember the farther something is away, like the stars, the farther back in time you are seeing.

It does not take much expense or effort to see this with a time predominate way of thinking. We are so used to a space predominance, that it biases us from seeing glimpses of more profound things which are right before us all along.

The nanoseconds of time difference are for practical purposes irrelevant, but if the distances become more of days, weeks, months or especially years, then we suddenly realize that what is so subtle in our everyday world is very profound. All around us in our human experience, this has been going on always. But it is thinking with a temporal predominance that allows us to 'see' far more than we have before. Society and even 'objective' science has biases, for humans are involved. These paradigms are what limits many to just the more that there is to measure and observe. And by observation mostly based on light, we can hear and see as well as smell. For events that are considered no more, or yet to be, in the Block Time of Einstein's work, are real. They for all practical purposes are no more kinetically interfering enough or 'relevant' to our everyday experiences.

One of my colleagues, David Morrison has asked the profound question, "How far away is tomorrow?" We can ask, 'how far away is yesterday?' as well. This can be answered by basing all events on the speed of light as our reference, which is the 'surface' of our superuniverse, around 300,000 kilometers/ second.

Take 12 noon on our present moment of 'today', our present now. Multiply the speed of light by 60 for a minute, then by 60 for an hour, then by 60 for a day of 24 hours. To shortcut this multiply the speed of light by 3600, then by 24. You now have the distance to tomorrow, and in the opposite, temporal direction to yesterday ... 25,920,000,000 kilometers away. This obviously equals one light day.

To find 'where' in space tomorrow is coming from, or where in space yesterday has gone to, we have the distance in space of an event in time. Now to get to this yesterday or tomorrow requires us to travel at 370,000 times c, with the loss of inertia and mass. That is what gives us kinetic interference in order for something to 'touch' or affect something else in the present as we understand it.

We now have to become tachyon-like for past, and anti-tachyon-like for future, while losing all of our tachyon, or slower than light-like attributes. This is what Hypergeometric Mechanics is all about. So at 370,000 times the speed of light, yesterday and tomorrow are still one light day away, but only *1.80 x 10 e -8*, which is repeating interestingly enough, and also equal to around 18 nanoseconds. This is the similar effect of 100 years ago or in the future, as 100 light years from now, but they are in opposite directions temporally; yet at 370,000 times light speed, each is just under 2½ hours away from the now of reference.

Get ready for the ride of your life! And though we do see always through a glass darkly as flesh and blood human beings, it is still so fantastic to glimpse what we are allowed to see.

Soon we must get you under way; allow me to answer the chapter's title question, 'Is future past, and is past future?' We can better say yes, but on opposite sides of the Superuniverse's Event Horizon. Each is tachyon world-like to the other. There is no kinetic interference between the two opposite flows of time, so as to render them practically irrelevant, yet with enough resolution possible glimpses of one another.

Welcome to the future ... and the past ... and the present!

Chapter Eight

Begin Your Journeys
Dear Time-Space Traveler

We need to go through the steps so that you can make practical what seems at first so theoretical, even impossible. In *HYPERSPHERE … A JOURNEY AT THE SPEED OF GEOMETRY* we introduced Hypergeometric Mechanics, which is dealing with the physics of very extreme states of geometry for matter.

In this work, we are actually considering the operating mechanisms that are utilizing such extremes. Now we get to personally experience from hypothesis and experimentation and limited human involvement.

First our power requirements are not as prohibitive as one may initially be concerned about. Yes hundreds of watts will be required, yet once inertia is overcome, much just continues, for we then hit this plateau where no more energy is needed. The maintaining of oscillation and the driver magnetic field is important, yet the drain of power soon becomes stabilized and not much current is drawn beyond just this 'cruising along' status.

There is radiation from the leakage as the Event Horizon approaches, so this is not for children, the pregnant or those wanting to bear children someday. A cheap Geiger-Muller Counter is fine to watch this flux, and if you really want to make your journeys, then you need to accept these conditions. Each end of the entanglement has considerable radiation from leakage; so going back or forward temporally will cause this.

The oscillations with the disc are less problematic, but the leakage source is the oscillating Hypersphere, the payload of our non-linear, harmonic, time-space conveyance. With the oscillations we are shaking the fabric of space-time and so what seems to be the case, this shaking of the local universe's foliation increases background radiation quite a bit.

To go back, the Hypersphere will appear to rise and flatten. With this is a lot of leakage and just when it begins to fade and seems transparent, is when the leakage lessons.

If you are inside the Hypersphere (and you can build one as large as you would like) then you are subject to radiation during the transit and while in entanglement. You are a tachyon particle within the Hypersphere (acting also as a tachyon particle) and your external environment is the passing of the Event Horizon, both sides of the superuniverse's surface of immense radiation and subatomic particles fluxing. The outside world is a bright and reddened, fading ring, and you and the Hypersphere are also a bright then reddened fading disc.

Let us now start with a small magnetic driver ring of one meter. The Hypersphere that will sit within and hover in the hole of this driver ring can be any size that will work and any mass that will draw the least amount of current, whilst shaking the local space-time enough to warrant progress.

Your kitchen table top is fine, and having even one disc shall be adventure enough. Have a camera ready too. In order to travel a day back and not change world lines requires the maintaining of entanglement. To travel a day forward, entanglement is already there, for the object is time dilating and just going faster into the forward than it was before. This all has to just be accepted, for we are still learning and therefore do not fully comprehend all this. To test this, perhaps place a clock on the table and one in the Hypersphere. A mechanical clock works great. Inside you could also have a viewport with an electronic clock and camera. Program them to photograph out the port during entanglement.

For more distant times, you will need additional time for transit, even though the effects at the tachyon state are at velocities of 370,000 times c. Remember in our dissertation in the last chapter, in order to travel one light day, you need to traverse backwards in time around 18 nanoseconds. It is so fast, yet so profound.

The picture that you are going to get is in space, a light day away, on the outskirts of our solar system. It will not be you sitting at your kitchen table the day before tomorrow's experiment. For you to do that, you would need to calculate the exact position of the Earth and the Sun in the courses through our home galaxy, and even our galaxy's position. So to photograph yourself at your kitchen table the day before would require a precise accuracy of calculation and setting the Hypersphere to go to that position in space, one light day ago. Therefore if you do not see something strange when taking your picture the first time of your kitchen table from the future, do not fret. It is very difficult to get all these positions accurately enough, but the potential is there.

What will most likely happen if you do get all things right, is that you may notice something, but when you least expect it and from an angle you do not know initially. If this was to be a success, then you have already noticed some occurrence just observable.

As long as entanglement is maintained the future traveler is not lost to a different world line. If entanglement is lost, then the future traveler, who is now past, just seems lost from the time-space continuum of your world line. For the future, now past, the traveler is in an alternate history never to return, and having just 'disappeared'! It is so similar to falling through the Event Horizon and lost from our part of space-time and our forward time course into the future.

To go into the future one essentially time dilates and depending on the Lorentz factor, tomorrow can vary from hours to minutes. To hit the tachyon's 370,000 times the speed of light point is not needed really, but if you can hit it, then you save vast amounts of electrical power.

My favorite journalistic text was to give account of the transit from 2010 to 1895, which is covered in *ODYSSEY OF THE AGES*. That was 115 light years and 115 years back in time in approximately 2.875 hours. To return to the future, it was the same forward, but all at 370,000 times light speed. All so counterintuitive, yet fascinating!

With the disc you can minimize all this to photon leakage and staying in your own time, while photographing through the entanglement tunnel a past or future scene. You need to allow the effects to envelope you and your equipment, whilst you remain in your proper time and space. Again you will probably get your first image of somewhere in space distance equal to the passage of time that the photon's source of that light (that is taking to 'leak through' the tunnel) to the distance that it has to normally travel on the surface of our superuniverse. Thus yesterday and tomorrow are one light day away in time and that is 26 billion kilometers away in space. Your entanglement tunnel traverses that in around 18 nanoseconds!

So most likely your photograph will consist of space at that distance yesterday or tomorrow. Time-space and space-time travel are inter-related.

I have wondered, if we speculate that our brains are also quantum and relativistic machines, we may already be able to do some form of this already. For some, God's universe is stranger and far more personal than we might care to imagine. For others it is solace. Temporalize space, as we so spatialize time. They inter-react.

Perhaps those Biblical prophets had more substance than our modern world gives them credit for. In Hypergeometric Mechanics we are just trying to understand it mechanistically. I often say, 'Pray before you use these machines.' Well, just pray and see what happens, if you have enough faith. Perchance we are coming all the way around to a perspective upon our origins that implies where we once were. It is written: 'if you have faith as a mustard seed, then you can tell this mountain to move.' Possibly all along much of this was far more attainable, but in our age of advanced technology, science and math, we may be really getting a glimpse of what was. Now we struggle to do it ourselves, but then we were given all this and much more than we valued.

The past reveals the future; the future reveals the past. If I have been wrong, please prove it; for otherwise you will be proving me right. I do desire sincere, repeated experiments and results outside of my own work. It may be that you will be the one, or amongst others that will acquire the more refined and greater understandings where I have only begun to tread. It may be you who will open doors to our universe, and thus our superuniverse of time and space, than we afore scientifically and seriously considered.

In my article, 'The Tachyon World' much is well explained about all that we are dealing with as to the place where and when entanglement, tunneling and a rest speed of 370,000 times the speed of light in reverse time occurs. If we are making sense of this realm, then it is required for us to travel through time and space.

Just for interest's sake, with all of this application of Hypergeometric Mechanics being touted herein, what are some more interrelated_findings of our explorations? What have we learned from the Hypersphere's transits already?

Within the distances so far ascertained beyond the slowness of rocketry, the local interstellar medium (and at points above and below the galaxy's disc) are very cold. The cosmic ray flux is greater above or from the direction generally from the Norther Galactic Pole (NGP). This flux is because of the Milky Way's motion of travel in the intergalactic space above it. The range of distance above and below our Galactic disc was about 1,000 light years each way. The temperature is almost 10° Kelvin. But this was not always the case, for some measurements approached 20° K because of hot plasma clouds passing through at more increased densities than at other times. It is still our home 'galactic environment', though therein are other essential elements in our Galaxy's winds, flux of gases, molecules, charged particles and radiations swirling and inhabiting.

Some of the flux north of our Galaxy's course may be from the Andromeda Galaxy (M-31) and even a little from the Triangulem Galaxy (M-33) coursing toward us as well. They are also sending out radiations, charged particles, molecules and gases that hit their respective escape velocities.

These areas are reached when entanglement is limited to a range of 1,000 to 2,000 light years of spatial distance. This of course requires hours and hours of Hypersphere oscillation time, and would be in the realm of power requirements, say for Israel or other high national infrastructures.

To reach a cleaner intergalactic range requires about 50,000 to 100,000 light years, and then you are getting a better sampling of the Large and Small Magellanic Clouds and again Andromeda and Triangulem streams of debris in the local intergalactic winds.

Going back to our local interstellar medium, we have between Alpha Centauri and our solar system gases, molecular clouds, particles and radiations. Between our sun and the stars of Alpha Centauri the temperature fluctuates much more depending upon the densities of this intervening material and electromagnetic changing environment. The temperature of this area can be from around 50 to 170° K to even as much as 400° K, but varies far more because of the greater inherent densities that are much more active in frequency over such smaller volumes of space.

This is the realm of interstellar space and is the more attainable arena for time-space experiments for the amateur. The power requirements are far less because of the much smaller masses and their times needed to transit through the local universe's mantle of the superuniverse. Days and weeks and months, even at 370,000 times c, with the constant need to power the oscillator, is for the amateur prohibitive.

For the traverse of a century or slightly longer, we are talking of just hours of transit for the Hypersphere and its internal sensors and possible occupants. This also is benefited by less arduous calculations of accuracy for positions of objects the last 100 years or so and the next 100 years or so. The distance of 100 years thus is just around 2½ hours away! Easy for the amateur to do and enjoy, and still keep a job and have an otherwise normal social life.

It is 'radio' where the signal is a mass and not just an electromagnetic wave. Let us try to better prepare ourselves for such a journey to 100 years ago; maybe to observe without interference, or better very little interference, of our great grandparents. We need to dress, talk and act like people in those days, so we need to study such. And 100 years ago is well chronicled within our local libraries for research to aid our endeavor. Consider the peculiarities of even our use of everyday language. In English 100 years ago, 'fancy' meant what today is spoken as 'sporting'.

Inside the Hypersphere we do experience radiation increases too and especially from cosmic rays' secondary radiation. This secondary radioactivity results from particles and energy emitted from the hull as primary cosmic rays and radiations interact with the Hypersphere's structural components.

Bismuth, and less so pyrolytic graphite, are what makes the Hypersphere induce a reverse magnetic field to the driver element, so oscillations can occur for half a cycle. The local Inertial Geometric or 'gravitational field' produces the other half of the cycle.

One has to just accept with the limits of economics and power production of the amateur, that radiation is a strong factor to contend with. Yet, even most of the governmental/military operators will have unavoidable radiation to deal with. During tunneling and entanglement, it seems very little radiation occurs, as now one is at rest velocity and tachyon-like at 370,000 times the speed of light.

One can travel inside the Hypersphere or just use it as an automated payload. The latter saving a lot of costs by keeping the human element at home and healthier.

Here are some more interesting observations and measurements I would like to mention, if you will allow me. There is a machine that was first theorized (actually by me) back in the 1960s. It consisted of a rotating mass in a noncircular, better Non-Keplerian, track with the plane of the track vertical. It was primarily nothing more than a mass accelerator and decelerator that forced a mass in a Non-Keplerian course. In the vertical it accelerated the device in opposition to the Inertial Geometric acceleration, thus resulting in the appearance of velocity upward or downward or to hover. Acceleration opposing acceleration appears as velocity to opposed velocity; when one is greater than the other, the greater is the dominant direction of motion. When done upon a small boat in a pool and the track was horizontal, it accelerated the boat in the direction of the non-circular area of the track. The mass had wheels to follow the track and was driven by an axle connected to it. There were in the beginning two opposing masses and increased by increments to eight. The revolutions per minute (rpm) of the track, when fast enough, made for quite a smooth propulsive effect.

Again, make this a vertical plane track and the non-circular geometry of the track aimed upward, this whole contraption could rise, hover and sink slowly or quite briskly, as a velocity offsetting another opposed velocity. The differentials could be very gentle and small, or rough and brisk. It even could serve as a 'crane' of sorts, thus an 'Inertial Crane' utilizing a mode of Inertial Buoyancy.

It was an effective method of just employing dynamic inertial accelerations to make use of the local Inertial Geometric. These speeds were not fast, more like a few kilometers per hour. But if kept in a linear trajectory, they were always accelerating to a next stable velocity.

It is not the idea of 'canceling' an Inertial Geometric, but making use of it. One cannot 'cancel' the Inertial Geometric at all, for there is always some residual even in the extremes of quantum and relativistic mechanics of particles. It is only when they pass through the Event Horizon into their respective Tachyon State or the Tachyon World that they are inertially and kinetically irrelevant for all practical purposes of our everyday experience.

All this is a form of Inertial Buoyancy, similarly with an orbit. In an orbit, the Inertial Geometric is not really cancelled, it is more equally balanced against. How important a concept and its definition is to our being able to adequately think about things to engineer from the Laws of Physics. As a matter of fact, in our respective space programs, when things are in orbit it is referred to as 'microgravity' and the conditions are thusly treated.

If you have a sailboat, rather than trying to 'cancel' the wind blowing against you, you track your sail to take advantage of it. This way you are using what at first appears an opposing force and utilize it now more wisely at the proper angle, causing your ship to successfully 'go against' the wind.

When the Hypersphere's magnetic driver's half cycle pulses more strongly per second, the Hypersphere appears to rise from the driver's seating hole while keeping all accelerations and velocities well into classical mechanics' states. Oscillating in this way, the induced opposing magnetic field does lift the Hypersphere, but it is in preparation to go into oscillations that will be considered relativistic. Then when it rises it will distort to a disc and redden, fade and rise. These are more optical distortions being seen from the real extreme Lorentzian effects happening to it, of approaching and passing through an Event Horizon.

We are doing similarly with making use of the Inertial Geometric, rather than trying to 'cancel' it. The 'substance' or 'something' that is the Inertial Geometric is for us mathematical and appears in applied mechanics. Yet we are very limited in that it specifically really is. We are faced with a concept limit is the best way to put it. All I understand is how to use it, like an ancient sailor who cannot really see that I am making use of flowing air molecules. It is written that 'we see through a glass darkly', and yet where the wind comes from or where it goes we do not know; we just know we can use our sails to traverse across a sea.

The manipulations we make upon the Hypersphere are articulated in its local space-time. It is played upon, so that it is as an instrument of music, responding to its extreme environment beyond what it would normally experience in an everyday and common state at rest. What a symphony!

Here we are seeing the most simple machine, not the wheel, but the pendulum. And a derivative of the pendulum is a spinning mass at the end of a tether. Now spinning at a certain radial velocity, the pendulum oscillates from rest to most kinetic state then rest, and back again. Each will dissipate its energy if not being continually pulsed to appreciate any such dispersing of friction or inertial resistance. It is such a primitive set of mechanisms that elude to grander schemes of matter, energy and motion, than at first ignored. The simpler a mechanism, the more basic the function being studied.

Both the pendulum and the spinning mass are replicating in function a simple structure in motion. There are profound implications that require tedious patience to observe and measure such, as to sense the far gentler manifestations of space-time and mass and motion. Approximate nanometers and millimeters are subtle functions to light years and centuries.

An oscillator begins the process and it has to span from millimeter to infrared. The most observed place of interesting action occurs in the Terahertz range. It has to be stable and it has to be amplitude modulated in order to articulate the driven mass, or payload, by the magnetic driver ring. The oscillations are sent into the ring through a magnetic amplifier. Here the pulsed magnetic field induces an opposing similar pulsed transient magnetic field from the Hypersphere. When the frequency is slowly increased then smoothly, like a musical instrument, it will rise some. Slowly the increased frequency and amplitude offset the increasing inertial resistance of the Hypersphere. Keeping the wavelength as small as possible, one can with practice, play this well and safely.

With properly pulsed harmonics and amplitude, one can jump the Hypersphere and the payload into a higher state enough to initialize a plateau where 'slippage' occurs. This is very close to the DeBroglie wavelength of the culmination of payload mass being articulated. It is very difficult to surmise where or when this occurs for the Hypersphere's space-time environment, for it is an 'area' not a specific place or time in its state of oscillation. Every 'flight' of the Hypersphere is different and unique to some extent, for the local mass being oscillated and the local curvature of space-time is not constant. The frequencies and amplitude required are thus slightly different. It is more art than science in this case.

As in the Quantum world, fuzziness is where and when you will be able to approximate; specifics seem less and less attainable. It is not for the impatient. Remember, you have needed to listen to the silence between the silences, and look for the shadow between the shadows already. Most science is done for the fastest and most profitable means, and so many have missed the intrigue and delight of it. So you, unrestrained have the luxury if one has such inclinations, to so tediously study finer resolutions of matter, energy, motion and space-time. Yet, for the amateur, it is a feasible and wonderful exploration. In most of the greatest discoveries in science, the lonely amateurs, not the large organizations, opened new doors for humankind.

I need to encourage you, even if many pursue, to prove me wrong. For in some ways I may be. But in your endeavors to do so, you may wind up trying to convince others of these shapes of things that have come, yet were always here. Please just give me fair credit ... that is all I ask in your attempts at all this. And no we are not finished with these instructions yet. You shall be one of the few beginning this journey into the still somewhat dark ages of time-space, opening doors to other times and worlds.

As the Hypersphere reaches extreme conditions per its local curvature, it then rises some and reddens, and seems to rise and fade more, becoming thin and disc-like. In a moment it is transparent and fades so away. It is now tunneling and as long as the driver ring continues to oscillate the non-visible Hypersphere, it is also entangled. As each approximate 1½ hours pass in your local time and space, the Hypersphere has passed approximately one light year in distance and one year back in time. Macroscopic scaled objects in such a state are now tachyon-like. They are also Quantum-like and of such a relativistic state that we are only allowed more uncertainty as Heisenberg's Uncertainty Principle whispers. We are now in a time and place of greater counterintuitive predominating our common notions of here and now, and space and time. Domination of time over space is when and where we are at a 'rest' velocity of 370,000 times the speed of light!

This 'slippage' of the Hypersphere in between the foliations of our local part of our understood universe, has allowed our mechanism to make a geometric chord below the speed of light surface of our local space-time. In this Tachyon World we find ourselves now, to our reference of space and time, dealing with time in reverse and distances on our surface to be much shortened. Now for this Tachyon World, we are the ones in reverse time (strange to those who may inhabit there) and under their surface of 'normal space-time'. We are indeed the strange ones. Here the two temporal, direction opposing sides of the superuniverse is manifest. Each side needs the other to exist. To look out your Hypersphere window, all that you would see is within the tunnel of entanglement. When you left your world, you appeared as a thin reddening disc and faded for an immeasurable moment, then reappeared. If you were to stop or better slow enough, you would see yourself in a universe with stars and galaxies similar to where you had just left. You would wonder if any of this had truly worked. But that is what is on the other side of the event horizon of Event Horizons, only the time direction is reversed.

Now as the programmed, if you were in the Hypersphere (but better if you first equalized the oscillations some) and you had precisely aimed at a distance of 100 light years away where Earth was, then you could have the Hypersphere produce an image, hopefully close-up. For discussion's sake, let's say this image was telescopic and over London or New York from a height of 100 miles. You would see what was going on in your appropriate time of 100 years ago. Granted this requires extreme precision. Also you should not be detected but even if you were, no one would be able to find any relevance to what they 'think' they may have seen. Thus your kinetic interference was thankfully minimal enough.

Think about this for a moment. If we understand that when we are looking into the night sky, into the past, the light at 100 years ago is 100 light years away; while knowing it is so weak and lacking definition, we take it for granted as a concept. Yet, what we have done is to go in reverse time and so now can observe the same close-up, but with greater strength and definition with our Hypersphere probe and its telescopic camera, while being so non-interfering as to not be a problem. Also we were able to cover 100 light years in space in nearly 2½ hours of our time and place of origin (your apartment and kitchen table).

Now to return you can just slow down the oscillations and the Hypersphere will return, for it was actually occupying two times and places at the same time and both places (whew!), and now fades in from bluish to regular light as it re-appears and fills back to a sphere from a disc in its return.

You retrieve your camera and view your well finished images, which required 370,000 times c to accomplish. You now rest, reflect upon all this and get back to work and your everyday world, for you need to make a living.

You contemplate something much bigger about your own traverse, but realize that for now, it is beyond some of your abilities, and much of your understanding. But then there are the discs we have incorporated as a sub-mechanism for the Hypersphere.

What do these discs with the two oscillating masses do? We shall delve into such shallower experiments in a later chapter.

Chapter Nine

Unbeknownst to Most

In 2010 information was sent to specific individuals and groups within the government and military of the United States, Great Britain and Israel. **The more powerful energy requirements and scientific accuracy and educational brain trusts already have much archived.** It has been Israel though, that seems to have the greatest need and has made possible use of Hypergeometric Mechanics because of her almost incessant necessity for defense and survival.

Using a ship or submarine (or even a combination of the two marine vessels) with adequate nuclear power sources, makes for an almost perfect support platform. Within the dedicated ship and/or submarine with appropriate engineering modifications, the launch, articulation and receiving of a Hypersphere system is as world-wide as are the Earth's oceans.

The capability to more privately operate such a system and to 'aim' to any part of the heaven's celestial sphere is total. The northern and southern targets for experiments of payloads light years away at 370,000 times the speed of light in attendant reverse time includes all known interstellar and near inter-galactic soundings. The required megawatts, especially for the modulation pulsed upon the standard continuous 'carrier' of the payload mass, is best upon an ocean vessel of large size.

The unidentified ship pictured next is merely an example of what might be considered. And for example, Haifa, Israel is an excellent port and the Mediterranean Sea has enough expanse for use requirements as needed.

Use of more than one Hypersphere is another mode of operation in which interference patterns could be utilized. Two would generate bi-directional, and three even more multi-directional oscillations in the local space-time fabric. No need to even launch a Hypersphere, for the oscillations would be more than adequate for offensive and defensive use. The purpose of entanglement and tunneling for such are not required in these situations.

In our experiments from 1998 to 2010, it was deep space propulsion with extreme reverse temporal effects and their human lifetime convenience durations out to hundreds to thousands of light years; those were the prime reasons in our goals. **No-linear propulsion, as opposed to then and still so prevalent linear propulsion methods were found to be far more effective for such transits for scientific**

exploration. By 2010, the primary experiments were over: for long space distances, the rocket was obsolete! The counterintuitive implications of relativistic mechanics and quantum mechanics begged to risk leaving behind linear sub-light considerations.

It was so easy to continue to assume that even at the unknown extreme geometries of physics in the realm of macroscopic mechanics, and that it was 'impossible', whilst not adequately tested enough. It is like the questions of antimatter falling as matter does, in a 'gravitational field', better Inertial Geometric. Still, at this writing, no one has really tested this thoroughly. We have known of the theory of antimatter since Paul Dirac and for decades after have found antimatter can even create minute amounts of it. Yet, still on and on, no one has taken the time to test this simple characteristic of antimatter. This must be assumed by those institutions that have made the great but limited understanding of what we do know about antimatter, unprofitable of further research or of low priority to pursue. These same ones will continue to practice the paradigms of linear propulsion as the 'only way to go' for spacecraft. Yes, the rocket and its velocities are fine for most of the solar system, but to the farther reaches of the solar system and the stars and the near space of our Milky Way Galaxy, we need something more profound.

If we can accept the proposition that time-space travel is operating in everyday environments as well as far distant ones, the cosmic scale, then these things are not so difficult to comprehend. We accept it already for the quantum scale, but for our macro-cosmic, everyday scale it is supposed to be 'not' possible. As I had mentioned earlier, reduce years and light years to nanometers and meters, and nanoseconds and seconds, then you can see that what is appearing so irrelevant practically, is far more real and 'everyday'. It is all a matter of enough resolution and thus perspective. It is also requires far more objectivity of scientific procedure, despite the uneasy ramifications that some of supposedly 'established' scientific theory, is more opinion than repeated verifiable evidence-based. Look at evolution and climate warming, as of this writing not all objective scientist can agree. Debating to prove God exists or not is convoluted in circular arguments with neither side, believers, agnostics nor atheists convincing the other. Has such irrational compassing repeatedly what objective science is in this 'post-modern' world? If religion is, as defined by Daniel Webster, a system of ideas, then 'objective' science is more personally biased religion; and without enough evidence either way, both sides are proving that to the other side. For believers see and find confirmation in the design of the Designer. The unbelievers see and find confirmation in the design of No Designer. Consider this, that God does not lie and neither do His rocks and stars! Also please consider, that with time humanity learns more and more, yet can never but approach its limit of knowing. Limits in mathematics are approached but never equalized or exceeded. If such is the case in objective mathematics, then why do rationalist and 'objective' scientists practice such irrational and un-objective behavior, without offering enough verifiable evidences?

Much of Hypergeometric Mechanics was implied by the action of the pendulum, which is the simplest machine. Spinning a pendulum leads to a circular course and function, not necessarily to the wheel. The wheel is not the simplest machine, though it is the simpler of so many.

As we have mentioned the proverbial ship, we can also mention the proverbial missile as a carrier for a small short-term operational Hypersphere, with a lesser pulse modulation to create a small and fast oscillation of the local space-time continuum. This is more tactical and similar to electronic warfare, but not such.

For example, we utilize a common form of cruise missile with waypoints which can be flown to a target very inconspicuously, until an element of surprise is needed. Now within its nose is the small Hypersphere to do its inertial undulations to affect the limited local area of concern. Because of its smaller size and adequate power generation onboard, its significance has to be very accurate for the small amount of time it has to operate sufficiently. Once it hits its pulsed modulation moment, then it does its powerful work. The mass itself can be just oscillated and pulsed, or it can be pulsed enough with a smaller mass for a total conversion that leaves no remains of the Hypersphere itself.

Mathematics as engineered to applied mechanics is very effective for time and space exploration and for defense of western civilization's foundations. Below is a rendition of a proposal for a cruise missile. Such is easy to use singly or in multiples for more directed inertial interference propagations. Again accuracy is imperative.

These are some of the more advanced means of Hypergeometric Mechanics than what the amateur would be able to afford to do or to support in their personal engineering support interface.

Better the use for humanity reaching for the stars and learning more about our local observable universe, especially the Milky Way environs, within and just above or below its disc structure.

This is an example of a cruise missile concept for small tactical theaters of defense, perhaps such as the Middle-East.

For to understand time, as part of the space-time continuum, is to understand that time-space is a very real construct having mechanistic function, though appearing counterintuitive initially. With further investigation, time and space make more sense, not for our convenience, but rather for our limited use of it to traverse great spatial distance, despite needing to experience the required relative temporal reverse durations. We may not like that requirement, but like the ocean and the atmosphere, we need to respect the requirements of their physics in order to survive and enjoy their rewards.

The world at large is a dangerous place. Seldom do great achievements not suffer scars from wars and civil unrest. Ancient ruins are such because of those human foibles just mentioned, and of course the normal degradation over time because of environmental changes and entropy.

It is a sad commentary upon human history, but precious things do need to be protected from those who are not capable of valuing such, or simply do not care. The protection of our western ideals needs to be kept vigilant. How it would be so wonderful if everyone of every generation would only continue to do their best; then the proper progress we could attain would allow us to reach for the stars and extend good will to all. But again, human history is the disappointing trail we have all left in our wake as individuals and generations. With the Hypersphere, even the lonely amateur can embark upon this new sea of exploration and become so infatuated with such possibilities, that perhaps one by one, all of us could produce sweeter and more lasting fruits.

Chapter Ten

The Use of the Discs

These simple discs are capable of such profound fruition. If there could be no easier short-cut, these discs are then sufficient. One disc is actually well enough. Two of them allows some more elaborate function, but such embellishments upon the singular function of one of them is not too forlorn.

Each disc fits into the palm of the average hand. It is initially a passive device, yet when it harmonically oscillates, it has a less powerful effect for far distance. Nearby is where it is of the utmost function. It seems to reinforce in a smaller volume of space, what the Hypersphere is affecting at a much greater distance. It seems to amplify the far smaller space with the same function as the Hypersphere's larger range. Thus holding one, you can either just sense the local space-time oscillations or effectively include a camera or sound recording device for some interesting experiential effects. Notably it is a very pronounced mechanism for the practical observation and measurement of Einstein's theory of Block Time. In Block Time, the past and future are as real as the present.

But as thankful as we are for the conceptualizations of theory, we so much desire to experience the actual active demonstration of all such. To see and hear, even smell or just touch, that of another time; that which the more distant temporally is also spatially.

Remember, 100 years ago is 100 light years away! Light speed is the inertial limit, our superuniverse's surface. That concept is quite counterintuitive already. Yet, we have just begun to consider far more inconveniences to come. We are not accustomed to such perspectives in our common everyday world, and some may just not like any of this beyond its being of academic meat to taste. But for the more daring, it is exploration that excites. So with respect to the extreme mechanics involved, then to ride these fringe environments may just be what one does want to do.

One of my fascinations is family history and perhaps one of yours as well. The desire not to interfere, but to just get a glimpse of a grandparent or great-great-great grandparent and his/her times. To somehow get some brief sense of the shape of those things to come, yet only glimpse a trend enough. I often say, one should pray to God before use of such mechanisms. It is written, to pray in all things. Here, for sure, you need to understand your potentials and your limits. You can only glimpse in most cases.

At 370,000 times the speed of light, you can traverse 100 years at a distance of 100 light years, in around 2 ½ hours one way. For you the direction of duration is one second per second and forward. It is what is relative to you that has the excitement of the ride you have chosen to make.
The accuracy required is immense to locate the required space 100 light years away, and may take several attempts, as well.

The illustration on the next page depicts the multiple use of the Hypersphere's inertial interference patterns generation (mentioned briefly in the last chapter). It can also help explain the discs in relationship to a single Hypersphere. Keep in mind the discs are resonant, thus in harmony with the Hypersphere. As such though they are not primarily generators, rather in their own parasitic oscillation are very much affecting a small local volume of their present space-time. As shown in the sketch, inertial interference (the 'shaking"' of the local space-time 'fabric') is directional and so able to be more effective for a camera to record and to actually observe with the naked eye.

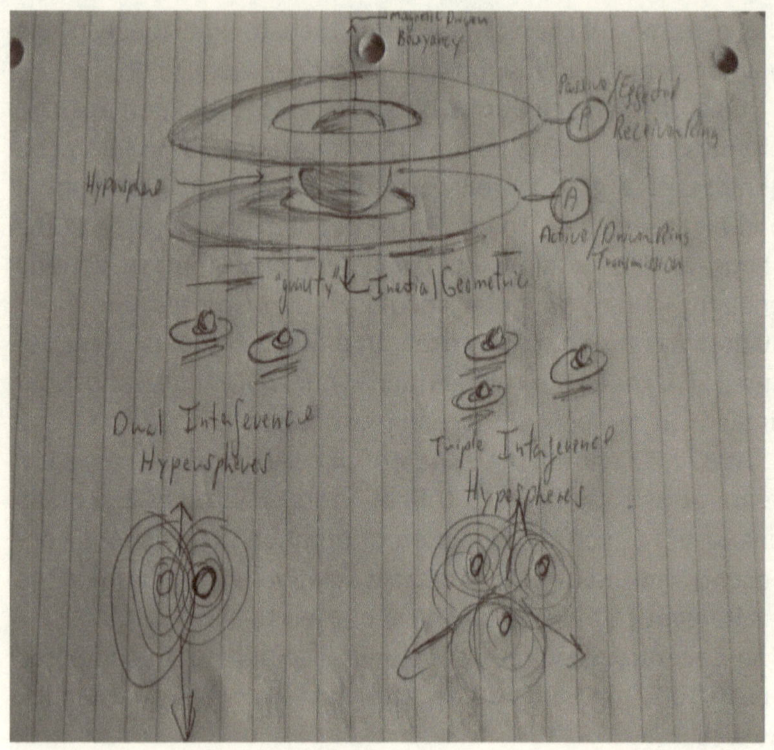

Your personal experience with these mechanisms of extreme geometry will better acquaint you with what is possible.

We have to get into angling of the Hypersphere to a spatial point, say 100 light years away for a 100 years forward or reverse temporal journey. At 370,000 times c, the Hypersphere only experiences around 2½ forward duration hours of transit. It is normal within the sphere for this experience, but if one could look outside the universe would be a disc-like glow around the Hypersphere. The higher the frequency, the thinner and more distinct the disc with red and blue sides alternating so fast. Then suddenly, one does not see any difference, as it just appears as a brilliant white disc.

This angling of the axial course of the Hypersphere is important in order to have it be where the Earth was 100 years ago. Spatially that is not far so far away. Consider that the Sun takes around 1,000 years to traverse a light year, therefore just one tenth of this distance is 100 light years. Now we have to also reckon the movement of our Galaxy and try to include the incongruities of all these movements of which we may not be aware. These discrepancies will include the average rotation of our superuniverse and our local observational universe floating upon it. But we need to counterintuitively consider the 'light speed surface' of 100 light years distance. This will hopefully make more sense as we progress. At 100 light years away, the light that left the particular place where the Earth and Sun were, is now arriving at our eyes.

Look into the night sky and the past, at various distances, is before you. The Orion Nebulae is about 1,500 light years away, so you are looking back 1,500 years. The star Alpha Centauri is around 4.32 light years away therefore you are looking back 4.32 years. The Andromeda Galaxy is approximately 2.5 million light years away consequently you are looking back around 2.5 million years ago. All this is based on the speed of light and our eyes receiving these photons from these emitters at the present moment of measurement and observation.

To observe our Earth and Sun at 100 years ago, we have a problem. We are at the photon emitter's source. Our photons are already gone away 100 light years from us. To travel into our own past, we have to short-cut the 100 light years to again receive the photons starting their transit and thus return to that time and space.

In order to do such, we have to go through the Event Horizon's local manifestation by articulating our Hypersphere properly, thus aiming it and oscillating it with the proper vector and duration. It would be very difficult to get 100 light years away by any inertial based system of propulsion, for we are ever striving to get close to the speed of light, but it is a mathematical limit we can never achieve. Rather to slip through an Event Horizon, traveling in reverse time duration at 370,000 times the speed of light, allows us to leave our 'world line', to tunnel and entangle in another 'world line' at its past point and reacquire photons emitted then. Not only that, being in this, our 'normal' forward time side of the Event Horizon, we are no longer even inertial here. We are rather at the past point, now inertial there. This means that we can kinetically interfere there, and only at the launch point somewhere, and when the Hypersphere left. Essentially, the Hypersphere is 100 years ago inertially and in the present only at its space above the driver ring, only mathematically. Better Non-inertially here and now, and 100 years ago inertially then and here!

Einstein's Block Time is vindicated, yet our own experience seems so counterintuitive in causality consequence, or certainly appears so. If one is not precise enough, then one may just entangle and tunnel to a point in space near the Earth and Sun. To that present population it would be irrelevant, for they actually would not have enough resolution to observe and the visiting Hypersphere. At most, early accurate attempts to explain would appear to anyone then as a moving object in the sky, but again irrelevant to most at that time.

Bear in mind, the Hypersphere is visiting time and space, so these two factors multiply complexity from just a time-transit perspective. If one is to accurately visit a time and place in the Earth's past, especially a populated area, extreme care must be taken to minimize kinetic interference. Then one surely has achieved much.

With just using the discs, much of this becomes more the observation and measurement of only leaked photons and some of the leaked acoustical-mechanical emanations. There is no kinetic interference. Amplification seems to help such emittances and so leakage is welcome.

If one were to entangle and tunnel through the volume of the discs, then possibly the world line would be changed and suddenly not be inertial here and now, only inertial then and there. But the discs would not prevent this. Because they are parasitic and not actually driving the oscillations, the discs are of very little insurance for security of a macroscopic object to reside on two world lines. To so swiftly be now on the opposite of the world line one was observing and measuring from, then one would now occupy in that past time and place their new 'present' and be gone permanently from their now and here of origin.

One has to learn to respect fire and the ocean as well as Hypergeometric Mechanics. The discs are a much safer way to study past time and place and to glimpse some future. But to glimpse the future, all one has to have is a high dilation of their present rate of time. The past is through the Tachyon World, which is how it would relate to us, the mantle of the superuniverse.

You cannot figure out which horse will win a race next week or next year. You can only accurately know what has been. You travel great distance into space by reverse time as a consequence of going 370,000 times the speed of light in a Non-inertial state. In so traveling, long distance in space is thus achieved with an appropriate reverse consequence of time.

It is counterintuitive to our common, everyday way of things. But at such a distance, any signal has been so minimized by the time it arrives here, and actually arrived before it left. Therefore those trying to measure it would not know when to or how to be able to detect it. Thus, causality is secure because of irrelevance of lack of resolution of observation and measurement.

The Inverse Square Law and the distance to and from Alpha Centauri are fine examples of all this. To get to Alpha Centauri, one has to be there after the 370,000 times light speed dissertations, 4.32 years ago from launch. Also flashing a laser beam to Earth then would not be very detectable after such a light speed journey. Similarly, the time of departure from Alpha Centauri and arrival on Earth would not be discernible because the exact moments are like Heisenberg's Uncertainty Principle, uncertain enough to not be measured and observed … though it happened, precisely forever. The future is by time dilation, and so sending back a signal from now to the past is based on spatial and temporal distance, requiring a return at 370,000 times c along the same world line.

Now we have an interesting more clear situation. Let us go 100 years into the future without reverse time, but through time dilation to get there to send a signal back. Then on a certain date and time in the future, send it to our associates in the here and now. To do this we will need to 'Hypersphere' back to this 'here and now', for that 'here and then' (a distance of 100 light years not just a spatial distance of one tenth of a light year). The direction is correct, but the temporal Non-inertial distance is still 100 light years for the Hypersphere's path, and it is then in the future to the receivers. This returning signal to the past is 100 years in the here and now for the receiver's future, at light speed. The Hypersphere sent to the future is now 100 years in that future sending its signal, restricted to the speed of light. In the future, the light signal will arrive in 100 years.

After 15 minutes in the future, the Hypersphere is able to come back to its original here and now, but it would have to go through the Event Horizon at 370,000 times the speed of light. But herein lies a problem, for the Hypersphere's drive disc is not only in the original here and now, additionally it has run for a mere 2½ hours; but for the drive disc, it has been 100 years. The drive unit to go into the future is quite the power drunk! It is not entangled to the future and has had to dilate the Hypersphere, whilst running externally all this time.

It is only through tunneling and entanglement that the Hypersphere can go back and forth from a future time-space to do what it does so marvelously. Yet again, causality is secure and God's Superuniverse as best we understand it, unfolds as it should. Consequently, tunneling and entanglement only work from future to past; it is strictly time dilation that runs from past to future. The only time that forward entanglement and tunneling apparently occur, requiring massive amounts of energy in a pulsed state, is when the Hypersphere tunnels and entangles forward. However there is a caveat when this is attempted ... the Hypersphere hits 370,000 times c, suddenly in reverse duration and into the past! So what appears as only apparent 'forward' entanglement and tunneling is actually 'reverse' entanglement and tunneling. This not only is evident with the Hypersphere in its experimentations, but with the simple tunnel diode of the 1950s.

Such future time dilation does occur, but is so energy prohibitive. In the Hypersphere it is well demonstrated that the driver has to continue running for the entire time to the future of years, whilst inside the Hypersphere it is such a forward dilated duration of maybe a few hours. And similarly, entanglement or tunneling appears at near light speed where there seems to be a place that going to the future is equivalent to 370,000 times the speed of light. Actuallythe Lorentzian effect is so pronounced that one is noticing within their Hypersphere one second per second, whilst outside the forward duration is 370,000 times their one second per second. So for every second one is dilating forward around 102.78 hours or 4.28 days. It is purely dilation but the effect is like tunneling. So far this is the best we can ascertain. Maybe in some way it is another form of tunneling and entanglement, a 'relativistic entanglement/tunneling'. Something yet to better fathom, and quite possibly you will be the one to figure this out.

To reinforce this consider that great distances in space are requiring far less energy, but the traverse is to the past with a temporal distance equivalent to the distance of a parallel light beam. What appears to be tunneling and entanglement are the particles going backwards in time. The occurrence for so long has been of such short distance of millimeters and nanometers, and maybe a few kilometers, that our resolution to observe and especially measure has been too inaccurate to understand this.

To glimpse may be all we shall be able to do, but implications do beg questions within an intelligent experimenter. As I have said before, 'enjoy the ride!' Perhaps in time we can better understand through our ruminations and reflections all that is happening.

To arrive at this theory before practice, it was required to increase the resolution of observation, especially measurement, by use of new instruments. These two new instruments were the Temporal Diffraction Grating and the Hyperplane.

After all, to travel effectively to great distances of space, using the most energy economical method, is through the past. Far more energy is required for linear rocket trajectories that try to force their way through time dilation into the future.

For deep space travel and agreeing to the reverse temporal ramifications, human traverse in such extreme physics allows for interstellar travel, even above and below our Galaxy's plane. The limit to all this is the human involvement. Perhaps automated probes could more easily make such far deeper journeys. It would be wondrous to really get a better structural map of our home Galaxy. Even the Magellanic Clouds and the area across our Galaxy's diameter would be feasible within approximately a third of an Earth year.

Conceivably in such flights of intrigue, we may discover our place considerably more in the Creator's Cosmos, and we can end irrational, circular arguments trying to convince one side by the other of religion and philosophy. I admire Sir Fred Hoyle, Freeman Dyson and by abstraction some, Crick and Watson for their finding evidence for design by a Designer. I too embrace such conclusions. Possibly this is how some can better 'choose this day whom they will serve', as it is written. In the extremes of science and mathematics, particularly physics and geometry, we are surely better able to test any fanciful presumptions.

As in a court of law, verifiable evidence, especially repeatable, increasingly confirms any hypothesis. Untested idealisms, blind and biased opinions refuse to be proven. It is so often that quotations of scientists in support of one side of an argument also conveniently ignore or try to hide the quotations of scientists that have evidence against such. Thus reoccurring irrational arguments in themselves discredit the modern rationalist's neurotic need to avoid challenge. It can rarely be the case for real intelligent discussion when a few objective, humble scientists that choose to dare to disagree, provide repeatable and verifiable evidence. So often, one advertises their degree and major in one field to vindicate and avoid disagreement.

I bring this up to have you recognize that whatever you discover has to be solid and enduring of such academic onslaughts. This is the requirement for real progress that has to be made without some politically correct researcher redefining science to their own delusions of grandeur. The hypothesis of climate warming and evolution should better be defined as climate change and Inherent Differentiation (IH), which have been mentioned in my other publications. Climate is changing, animal and plant species are diversified, and these mechanics still require far longer periods of time to ascertain than what is convenient to tyrannically badger true thinkers to 'just accept' because they say they are enlightened.

When a true scientific mind does not have enough evidence that can be verified, one just has an interesting hypothesis, until proven. A wiser choice is to wait and be patient, and even dare to admit that they do not know. 'Sic simper tyrannous' to the idealist bullies that cannot prove by practice and experience.

A word of advice, whilst they spout and bully you and your allies: in time, you own evidence that will speak for itself. In the meantime, enjoy what they cannot bother to understand. It is a very lonely trek and uncomfortable for the true explorer. Do not be distracted by the spoiled arrogance or the slights of the rational and enlightened. Journey into the realms of extreme physics anyway. As it is written: 'Let the dead bury their dead', whilst you go to the stars!

THE FOUR STEPS OF A
<u>NEW IDEA</u>

1. FIRST, THEY IGNORE YOU
2. THEN, THEY LAUGH AT YOU
3. THEN, THEY WANT TO FIGHT YOU
4. THEN, YOU WIN

Mahatma Ghandi

It is very fascinating to read what Mahatma Ghandi stated above, and what Albert Einstein suggested below.

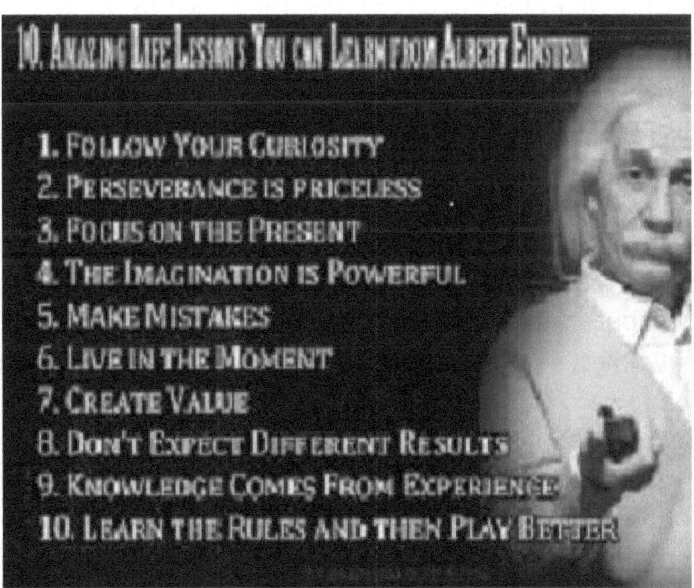

With Ghandi and Einstein, we have two intelligent and caring individuals that had to surely go against the common flow of locked opinions and thoughts of the so many who cared not to listen and understand. It took the vindication of unfolded experience, and time of history and science that today makes them some of our heroes. These who inspire had also walked the same path you now tread.

Many are sensing that we are already at the start of World War III. On March 3, 2015, Prime Minister of Israel, Benjamin Netanyahu spoke before the United States Congress. It was a solemn and moving speech that emphasized the dangers of a nuclear Iran to the rest of the world. In these days that seem to be nearing the end of civilization, hopes dim for another generation of thinkers, dreamers, visionaries and inventors trying to make a much better world; it is only for the courageous to go on. You may have such new ideas or just inklings of fuller imaginations of things to come. You may have even been blessed with enough rejection and deflation to fight alone and away from others for your own shape of some things yet to come.

Similar to the late nineteenth century, as for Faraday, Maxwell and others, there is a time still for the lone experimenter and inventor. Here is a vast field so barely glimpsed and pregnant with such profound possibilities.

You will need to take detailed notes, measure and re-measure. You will suffer burns and contusions and even become ill to some extent from what you may allow yourself to be exposed to. It may cost you any chance of a social life, and at times friends. But to persist, try again and carry on will be reward for your time and efforts.

Get the photographs, audio recordings and data, and then take the reflective time to digest all this you are exploring. There were times that it would take hours or days to re-orient myself from such similar pursuits.

Hypergeometric Mechanics is so on the edge and extreme that it is far too counterintuitive for many in the professional areas of math and science to understand. This was the same with relativity and quantum mechanics for decades.

You may find yourself redefining and thus developing new concepts to what you are trying to fathom. For some of you this will bring on new discoveries. It is daring to try new things and test new paradigms. Where others have not or will not, you, as I, will.

Launching

Sadly, our known history is witness to the dissipation of so much of humanity's achievements, and so only rare accomplishments survive in the long term. Now we are again in such dire straits. Wars and violent civil unrest are here and there. Culturally we are digressing more and more. Refinements and advancements in arts, science and technology grow in one quarter, becoming stagnate and even lost in another. The empire builders are able to so quickly, powerfully and quietly usurp what all of civilization deserves ... stealing and utilizing all such for their own self-gain. It is as though another Dark Ages is slinking upon humanity, whilst any advancements are so swiftly harnessed for a powerful few.

To get humankind to the stars and to better explore all the dimensions of time and space, advancements cannot be so hindered by elitist. That is why I have written this handbook.

There is so much more to all that we experience in our world and other worlds. This Earth is a wonderful planet and our ecosystem is easily Galactic in extent and requirement. Even extra-galactic is essential. Our present of time and space seems so magnified, and it is absolutely significant. But beyond our now and here is so much more that sustains it ... from deep time past, to deep time future, and from so very near to deep space.

What you are exploring is but a very minute spectrum of the time-space continuum. To a human lifetime, 100 years ago and 100 years hence are so dramatic in scale. To go to the Moon or Mars is a wonderful measure of exploration in our solar system. But from the minutest of scale to the grandest of scales, we are but a very small and humble ingredient, sometime and somewhere in-between.

Thank you for your inquisitiveness and daring to wonder. I often say so many adults just think, whilst children wonder. Be that child still within to wonder. Let your imagination inflame your thinking and experimentation to try to glimpse yesterday, now and tomorrow ... and from here to light years away.

As in H G Well's THE TIME MACHINE, the 'Time Traveler' dares to push that lever forward, at first gingerly I am sure, then with more mustered courage, farther and more firmly. Slowly around him the consequence of his premeditated act appears. Now the Sun and Moon blur in their paths through the sky as slightly shifting arcs upon a grey, pulsating day-night cycle in the heavens smeared together above. Soon the seasons more and more swiftly pass, and then onward does the traveller, with extreme wonder, journey until he arrives in the year 802,701 AD!

Again, I welcome you to the future, the past and the present. See you in the future!

- THE TIME SPACE TRAVELER, 2014 AD